DATA MINING AND DATA WAREHOUSING

Kanika Lakhani Chaudhary

P.hd (Pursuing), M.Tech (CSE)
Assistant Professor
Manav Rachna College of Engineering, Faridabad

Gaurav Girdhar

Assistant System Engineer
Tata Consultancy Services

S.K. KATARIA & SONS®

Publisher of Engineering Books

4885/109, Prakash Mahal, Dr. Subhash Bhargav Lane,
Opp. Delhi Medical Association, Daryaganj, New Delhi-110002 (India)
Tel.: +91-11-43551243; Mobile: +91-9871775858, 7678425209
E-mail: katariabook@gmail.com, katariabooks@yahoo.com
Website: www.skkatariaandsons.com
(e-books, subscriptions and bulk buying also available on our website)

Published By :

S.K. KATARIA & SONS®
4885/109, Prakash Mahal, Dr. Subhash Bhargav Lane,
Daryaganj, New Delhi-110002 (India)
Tel.: +91-11-43551243; Mobile: +91-9871775858, 7678425209
e-mail: katariabook@gmail.com, katariabooks@yahoo.com

First Edition: August, 2012, Reprint: 2025

Price: ₹ 395/-

Laser Typesetting at: Goswami Associates, Delhi.

Printed at: Arora Printing Press, Delhi-110051

Preface

This book is meant for engineering students of UP Technical University and Uttrakhand Technical University. This book has been written in accordance with the updated syllabus for B.Tech. Course with llustrative figures. The aim is to provide an up-to-date account of relavant topics. Due consideration has been taken to make the language easy to understand that facilitates easy understanding of the concepts in each chapter.

Kanika Lakhani
Gaurav Girdhar

Acknowledgement

We are thankful to Almighty God for blessing us with the strength to face any situation with courage and patience. We extend my thanks to our families who encouraged us to finish the task successfully on time. We also thank our friends and faculties for their inspiration and blessings. We convey our thanks to Mr. Sanjeev Kataria, the publisher of this book, Mr. Sharda Singh and his entire team for their continuous interaction and suggestions for improvement in the manuscript and bringing out this book in time.

Kanika Lakhani
Gaurav Girdhar

Dedicated to
My Husband

Contents

3 Associative Mining Rule 36–62

4 Classification and Prediction 63–86

7 Data Warehousing and OLAP 144–163

8 Client-Server Computing 164–183

9 Web Mining 184–202

10 Security in DWH and Data Mining 203–225

Appendix 226–246

Index 247–251

CHAPTER 1

Data Mining

Data mining is the process of analyzing data from different perspectives and summarizing it into useful information—information that can be used to increase revenue, cuts costs, or both. Data mining software is one of a number of analytical tools for analyzing data. It allows users to analyze data from many different dimensions or angles, categorize it, and summarize the relationships identified. Technically, data mining is the process of finding correlations or patterns among dozens of fields in large relational databases.

Although data mining is a relatively new term, the technology is not. Companies have used powerful computers to shift through volumes of supermarket scanner data and analyze market research reports for years. However, continuous innovations in computer processing power, disk storage, and statistical software are dramatically increasing the accuracy of analysis while driving down the cost.

1.1 INTRODUCTION TO DATA MINING

Data mining is primarily used today by companies with a strong consumer focus - retail, financial, communication, and marketing organizations. It enables these companies to determine relationships among "internal" factors such as price, product positioning, or staff skills, and "external" factors such as economic indicators, competition, and customer demographics. And, it enables them to determine the impact on sales, customer satisfaction, and corporate profits. Finally, it enables them to "drill down" into summary information to view detail transactional data.

With data mining, a retailer could use point-of-sale records of customer purchases to send targeted promotions based on an individual's purchase history. By mining demographic data from comment or warranty cards, the retailer could develop products and promotions to appeal to specific customer segments.

While large-scale information technology has been evolving separate transaction and analytical systems, data mining provides the link between the two. Data mining software analyzes relationships and patterns in stored transaction data based on open-ended user queries. Several types of analytical software are available: statistical, machine learning, and neural networks. Generally, any of four types of relationships are sought:

- **Classes:** Stored data is used to locate data in predetermined groups. For example, a restaurant chain could mine customer purchase data to determine when customers visit and what they typically order. This information could be used to increase traffic by having daily specials.

- **Clusters:** Data items are grouped according to logical relationships or consumer preferences. For example, data can be mined to identify market segments or consumer affinities.
- **Associations:** Data can be mined to identify associations. The beer-diaper example is an example of associative mining.
- **Sequential patterns:** Data is mined to anticipate behavior patterns and trends. For example, an outdoor equipment retailer could predict the likelihood of a backpack being purchased based on a consumer's purchase of sleeping bags and hiking shoes.

Data mining consists of five major elements:

- Extract, transform, and load transaction data onto the data warehouse system.
- Store and manage the data in a multidimensional database system.
- Provide data access to business analysts and information technology professionals.
- Analyze the data by application software.
- Present the data in a useful format, such as a graph or table.

Different levels of analysis are available:

- **Artificial neural networks:** Non-linear predictive models that learn through training and resemble biological neural networks in structure.
- **Genetic algorithms:** Optimization techniques that use processes such as genetic combination, mutation, and natural selection in a design based on the concepts of natural evolution.
- **Decision trees:** Tree-shaped structures that represent sets of decisions. These decisions generate rules for the classification of a dataset. Specific decision tree methods include Classification and Regression Trees (CART) and Chi Square Automatic Interaction Detection (CHAID) . CART and CHAID are decision tree techniques used for classification of a dataset. They provide a set of rules that you can apply to a new (unclassified) dataset to predict which records will have a given outcome. CART segments a dataset by creating 2-way splits while CHAID segments using chi square tests to create multi-way splits. CART typically requires less data preparation than CHAID.
- **Nearest neighbor method:** A technique that classifies each record in a dataset based on a combination of the classes of the *k* record(s) most similar to it in a historical dataset (where *k* 1). Sometimes called the *k*-nearest neighbor technique.
- **Rule induction:** The extraction of useful if-then rules from data based on statistical significance.
- **Data visualization:** The visual interpretation of complex relationships in multidimensional data. Graphics tools are used to illustrate data relationships.

1.2 KNOWLEDGE DISCOVERY

With the enormous amount of data stored in files, databases, and other repositories, it is increasingly important, if not necessary, to develop powerful means for analysis and perhaps interpretation of such data and for the extraction of interesting knowledge that could help in decision-making.

Data Mining, also popularly known as *Knowledge Discovery in Databases* (KDD), refers to the nontrivial extraction of implicit, previously unknown and potentially useful information from data in databases. While data mining and knowledge discovery in databases (or KDD) are frequently treated as synonyms, data mining is actually part of the knowledge discovery process. Figure 1 shows data mining as a step in an iterative knowledge discovery process.

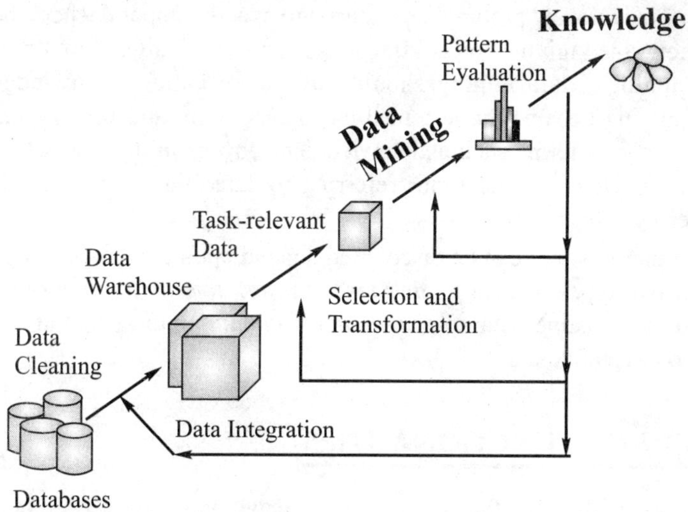

Figure 1.1. Data mining is the core of knowledge discovery process

The Knowledge Discovery in Databases process comprises of a few steps leading from raw data collections to some form of new knowledge. The iterative process consists of the following steps:

- **Data cleaning:** also known as data cleansing; it is a phase in which noise data and irrelevant data are removed from the collection.
- **Data integration:** at this stage, multiple data sources, often heterogeneous, may be combined in a common source.
- **Data selection:** at this step, the data relevant to the analysis is decided on and retrieved from the data collection.
- **Data transformation:** also known as data consolidation, it is a phase in which the selected data is transformed into forms appropriate for the mining procedure.
- **Data mining:** it is the crucial step in which clever techniques are applied to extract patterns potentially useful.
- **Pattern evaluation:** in this step, strictly interesting patterns representing knowledge are identified based on given measures.
- **Knowledge representation:** is the final phase in which the discovered knowledge is visually represented to the user. This essential step uses visualization techniques to help users understand and interpret the data mining results.

It is common to combine some of these steps together. For instance, *data cleaning* and *data integration* can be performed together as a pre-processing phase to generate a data warehouse. *Data selection* and *data transformation* can also be combined where the consolidation of the data is the result of the selection, or, as for the case of data warehouses, the selection is done on transformed data.

KDD is an iterative process. Once the discovered knowledge is presented to the user, the evaluation measures can be enhanced, the mining can be further refined, new data can be selected or further transformed, or new data sources can be integrated, in order to get different, more appropriate results.

Data mining derives its name from the similarities between searching for valuable information in a large database and mining rocks for a vein of valuable ore. Both imply either sifting through a large amount of material or ingeniously probing the material to exactly pinpoint where the values reside. It is, however, a misnomer, since mining for gold in rocks is usually called "gold mining" and not "rock mining", thus by analogy, data mining should have been called "knowledge mining" instead. Nevertheless, data mining became the accepted customary term, and very rapidly a trend that even overshadowed more general terms such as knowledge discovery in databases (KDD) that describe a more complete process. Other similar terms referring to data mining are: data dredging, knowledge extraction and pattern discovery.

The kinds of patterns that can be discovered depend upon the data mining tasks employed. By and large, there are two types of data mining tasks: *descriptive data mining* tasks that describe the general properties of the existing data, and *predictive data mining* tasks that attempt to do predictions based on inference on available data.

1.3 DATA MINING FUNCTIONALITIES

The data mining functionalities and the variety of knowledge they discover are briefly presented in the following list:

- **Characterization:** Data characterization is a summarization of general features of objects in a target class, and produces what is called *characteristic rules*. The data relevant to a user-specified class are normally retrieved by a database query and run through a summarization module to extract the essence of the data at different levels of abstractions. For example, one may want to characterize the 'OurVideoStore' customers who regularly rent more than 30 movies a year. With concept hierarchies on the attributes describing the target class, the *attribute-oriented induction* method can be used, for example, to carry out data summarization. Note that with a data cube containing summarization of data, simple OLAP operations fit the purpose of data characterization.
- **Discrimination:** Data discrimination produces what are called *discriminant rules* and is basically the comparison of the general features of objects between two classes referred to as the *target class* and the *contrasting class*. For example, one may want to compare the general characteristics of the customers who rented more than 30 movies in the last year with those whose rental account is lower than 5. The techniques used for data discrimination are very similar to the techniques used for data characterization with the exception that data discrimination results include comparative measures.
- **Association analysis:** Association analysis is the discovery of what are commonly called *association rules*. It studies the frequency of items occurring together in transactional databases, and based on a threshold called *support*, identifies the frequent item sets. Another threshold, *confidence*, which is the conditional probability than an item appears in a transaction when another item appears, is used to pinpoint association rules. Association analysis is commonly used for market basket analysis. For example, it could be useful for the 'OurVideoStore' manager to know what movies are often rented together or if there is a relationship between renting a certain type of movies and buying popcorn or pop. The discovered association rules are of the form: $P \rightarrow Q$ $[s, c]$, where P and Q are conjunctions of attribute value-pairs, and s (for support) is the probability that P and Q appear together in a transaction and c (for confidence) is the conditional probability that Q appears in a transaction when P is present.

For example, the hypothetic association rule: *RentType(X, "game") AND Age(X, "13-19") → Buys(X, "pop")* [*s=2% ,c=55%*] would indicate that 2% of the transactions considered are of customers aged between 13 and 19 who are renting a game and buying a pop, and that there is a certainty of 55% that teenage customers who rent a game also buy pop.

- **Classification:** Classification analysis is the organization of data in given classes. Also known as *supervised classification*, the classification uses gives class labels to order the objects in the data collection. Classification approaches normally use a *training set* where all objects are already associated with known class labels. The classification algorithm learns from the training set and builds a model. The model is used to classify new objects. For example, after starting a credit policy, the 'OurVideoStore' managers could analyze the customers behaviours vis-à-vis their credit, and label accordingly the customers who received credits with three possible labels "safe", "risky" and "very risky". The classification analysis would generate a model that could be used to either accept or reject credit requests in the future.
- **Prediction:** Prediction has attracted considerable attention given the potential implications of successful forecasting in a business context. There are two major types of predictions: one can either try to predict some unavailable data values or pending trends, or predict a class label for some data. The latter is tied to classification. Once a classification model is built based on a training set, the class label of an object can be foreseen based on the attribute values of the object and the attribute values of the classes. Prediction is however more often referred to the forecast of missing numerical values, or increase/ decrease trends in time related data. The major idea is to use a large number of past values to consider probable future values.
- **Clustering:** Similar to classification, clustering is the organization of data in classes. However, unlike classification, in clustering, class labels are unknown and it is up to the clustering algorithm to discover acceptable classes. Clustering is also called *unsupervised classification*, because the classification is not dictated by given class labels. There are many clustering approaches all based on the principle of maximizing the similarity between objects in a same class (*intra-class similarity*) and minimizing the similarity between objects of different classes (*inter-class similarity*).
- **Outlier analysis:** Outliers are data elements that cannot be grouped in a given class or cluster. Also known as *exceptions* or *surprises*, they are often very important to identify. While outliers can be considered noise and discarded in some applications, they can reveal important knowledge in other domains, and thus can be very significant and their analysis valuable.
- **Evolution and deviation analysis:** Evolution and deviation analysis pertain to the study of time related data that changes in time. Evolution analysis models evolutionary trends in data, which consent to characterizing, comparing, classifying or clustering of time related data. Deviation analysis, on the other hand, considers differences between measured values and expected values, and attempts to find the cause of the deviations from the anticipated values.

It is common that users do not have a clear idea of the kind of patterns they can discover or need to discover from the data at hand. It is therefore important to have a versatile and inclusive data mining system that allows the discovery of different kinds of knowledge and at different levels of abstraction. This also makes interactivity an important attribute of a data mining system.

1.4 DATA MINING SYSTEM CHARACTERIZATION AND ITS ISSUES

We have been collecting a myriad of data, from simple numerical measurements and text documents, to more complex information such as spatial data, multimedia channels, and hypertext documents. Here is a non-exclusive list of a variety of information collected in digital form in databases and in flat files.

- **Business transactions:** Every transaction in the business industry is often "memorized" for perpetuity. Such transactions are usually time related and can be inter-business deals such as purchases, exchanges, banking, stock, etc., or intra-business operations such as management of in-house wares and assets. Large department stores, for example, thanks to the widespread use of bar codes, store millions of transactions daily representing often terabytes of data. Storage space is not the major problem, as the price of hard disks is continuously dropping, but the effective use of the data in a reasonable time frame for competitive decision-making is definitely the most important problem to solve for businesses that struggle to survive in a highly competitive world.
- **Scientific data:** Whether in a Swiss nuclear accelerator laboratory counting particles, in the Canadian forest studying readings from a grizzly bear radio collar, on a South Pole iceberg gathering data about oceanic activity, or in an American university investigating human psychology, our society is amassing colossal amounts of scientific data that need to be analyzed. Unfortunately, we can capture and store more new data faster than we can analyze the old data already accumulated.
- **Medical and personal data:** From government census to personnel and customer files, very large collections of information are continuously gathered about individuals and groups. Governments, companies and organizations such as hospitals, are stockpiling very important quantities of personal data to help them manage human resources, better understand a market, or simply assist clients. Regardless of the privacy issues this type of data often reveals, this information is collected, used and even shared. When correlated with other data this information can shed light on customer behaviour and the like.
- **Surveillance video and pictures:** With the amazing collapse of video camera prices, video cameras are becoming ubiquitous. Video tapes from surveillance cameras are usually recycled and thus the content is lost. However, there is a tendency today to store the tapes and even digitize them for future use and analysis.
- **Satellite sensing:** There is a countless number of satellites around the globe: some are geo-stationary above a region, and some are orbiting around the Earth, but all are sending a non-stop stream of data to the surface. NASA, which controls a large number of satellites, receives more data every second than what all NASA researchers and engineers can cope with. Many satellite pictures and data are made public as soon as they are received in the hopes that other researchers can analyze them.
- **Games:** Our society is collecting a tremendous amount of data and statistics about games, players and athletes. From hockey scores, basketball passes and car-racing lapses, to swimming times, boxers pushes and chess positions, all the data are stored. Commentators and journalists are using this information for reporting, but trainers and athletes would want to exploit this data to improve performance and better understand opponents.
- **Digital media:** The proliferation of cheap scanners, desktop video cameras and digital cameras is one of the causes of the explosion in digital media repositories. In addition, many radio stations, television channels and film studios are digitizing their audio and video collections

to improve the management of their multimedia assets. Associations such as the NHL and the NBA have already started converting their huge game collection into digital forms.

- **CAD and Software engineering data:** There are a multitude of Computer Assisted Design (CAD) systems for architects to design buildings or engineers to conceive system components or circuits. These systems are generating a tremendous amount of data. Moreover, software engineering is a source of considerable similar data with code, function libraries, objects, etc., which need powerful tools for management and maintenance.
- **Virtual Worlds:** There are many applications making use of three-dimensional virtual spaces. These spaces and the objects they contain are described with special languages such as VRML. Ideally, these virtual spaces are described in such a way that they can share objects and places. There is a remarkable amount of virtual reality object and space repositories available. Management of these repositories as well as content-based search and retrieval from these repositories are still research issues, while the size of the collections continues to grow.
- **Text reports and memos (e-mail messages):** Most of the communications within and between companies or research organizations or even private people, are based on reports and memos in textual forms often exchanged by e-mail. These messages are regularly stored in digital form for future use and reference creating formidable digital libraries.
- **The World Wide Web repositories:** Since the inception of the World Wide Web in 1993, documents of all sorts of formats, content and description have been collected and inter-connected with hyperlinks making it the largest repository of data ever built. Despite its dynamic and unstructured nature, its heterogeneous characteristic, and its very often redundancy and inconsistency, the World Wide Web is the most important data collection regularly used for reference because of the broad variety of topics covered and the infinite contributions of resources and publishers. Many believe that the World Wide Web will become the compilation of human knowledge.

In principle, data mining is not specific to one type of media or data. Data mining should be applicable to any kind of information repository. However, algorithms and approaches may differ when applied to different types of data. Indeed, the challenges presented by different types of data vary significantly. Data mining is being put into use and studied for databases, including relational databases, object-relational databases and object-oriented databases, data warehouses, transactional databases, unstructured and semi-structured repositories such as the World Wide Web, advanced databases such as spatial databases, multimedia databases, time-series databases and textual databases, and even flat files. Here are some examples in more detail:

- **Flat files:** Flat files are actually the most common data source for data mining algorithms, especially at the research level. Flat files are simple data files in text or binary format with a structure known by the data mining algorithm to be applied. The data in these files can be transactions, time-series data, scientific measurements, etc.
- **Relational Databases:** Briefly, a relational database consists of a set of tables containing either values of entity attributes, or values of attributes from entity relationships. Tables have columns and rows, where columns represent attributes and rows represent tuples. A tuple in a relational table corresponds to either an object or a relationship between objects and is identified by a set of attribute values representing a unique key.

- **Data Warehouses:** A data warehouse as a storehouse, is a repository of data collected from multiple data sources (often heterogeneous) and is intended to be used as a whole under the same unified schema. A data warehouse gives the option to analyze data from different sources under the same roof. Let us suppose that 'OurVideoStore' becomes a franchise in North America. Many video stores belonging to 'OurVideoStore' company may have different databases and different structures. If the executive of the company wants to access the data from all stores for strategic decision-making, future direction, marketing, etc., it would be more appropriate to store all the data in one site with a homogeneous structure that allows interactive analysis. In other words, data from the different stores would be loaded, cleaned, transformed and integrated together. To facilitate decision-making and multi-dimensional views, data warehouses are usually modeled by a multi-dimensional data structure.
- **Transaction Databases:** A transaction database is a set of records representing transactions, each with a time stamp, an identifier and a set of items. Associated with the transaction files could also be descriptive data for the items.
- **Multimedia Databases:** Multimedia databases include video, images, audio and text media. They can be stored on extended object-relational or object-oriented databases, or simply on a file system. Multimedia is characterized by its high dimensionality, which makes data mining even more challenging. Data mining from multimedia repositories may require computer vision, computer graphics, image interpretation, and natural language processing methodologies.
- **Spatial Databases:** Spatial databases are databases that, in addition to usual data, store geographical information like maps, and global or regional positioning. Such spatial databases present new challenges to data mining algorithms.
- **Time-Series Databases:** Time-series databases contain time related data such as stock market data or logged activities. These databases usually have a continuous flow of new data coming in, which sometimes causes the need for a challenging real time analysis. Data mining in such databases commonly includes the study of trends and correlations between evolutions of different variables, as well as the prediction of trends and movements of the variables in time.
- **World Wide Web:** The World Wide Web is the most heterogeneous and dynamic repository available. A very large number of authors and publishers are continuously contributing to its growth and metamorphosis, and a massive number of users are accessing its resources daily. Data in the World Wide Web is organized in inter-connected documents. These documents can be text, audio, video, raw data, and even applications. Conceptually, the World Wide Web is comprised of three major components: The content of the Web, which encompasses documents available; the structure of the Web, which covers the hyperlinks and the relationships between documents; and the usage of the web, describing how and when the resources are accessed. A fourth dimension can be added relating the dynamic nature or evolution of the documents. Data mining in the World Wide Web, or web mining, tries to address all these issues and is often divided into web content mining, web structure mining and web usage mining.

1.5 CLASSIFICATION OF DATA MINING SYSTEMS

There are many data mining systems available or being developed. Some are specialized systems dedicated to a given data source or are confined to limited data mining functionalities, other are more versatile and comprehensive. Data mining systems can be categorized according to various criteria among other classification are the following:

- **Classification according to the type of data source mined:** this classification categorizes data mining systems according to the type of data handled such as spatial data, multimedia data, time-series data, text data, World Wide Web, etc.
- **Classification according to the data model drawn on:** this classification categorizes data mining systems based on the data model involved such as relational database, object-oriented database, data warehouse, transactional, etc.
- **Classification according to the king of knowledge discovered:** this classification categorizes data mining systems based on the kind of knowledge discovered or data mining functionalities, such as characterization, discrimination, association, classification, clustering, etc. Some systems tend to be comprehensive systems offering several data mining functionalities together.
- **Classification according to mining techniques used:** Data mining systems employ and provide different techniques. This classification categorizes data mining systems according to the data analysis approach used such as machine learning, neural networks, genetic algorithms, statistics, visualization, database-oriented or data warehouse-oriented, etc. The classification can also take into account the degree of user interaction involved in the data mining process such as query-driven systems, interactive exploratory systems, or autonomous systems. A comprehensive system would provide a wide variety of data mining techniques to fit different situations and options, and offer different degrees of user interaction.

1.6 SCOPE OF DATA MINING

Data mining derives its name from the similarities between searching for valuable business information in a large database — for example, finding linked products in gigabytes of stored scanner data — and mining a mountain for a vein of valuable ore. Both processes require either sifting through an immense amount of material, or intelligently probing it to find exactly where the value resides. Given databases of sufficient size and quality, data mining technology can generate new business opportunities by providing these capabilities:

- **Automated prediction of trends and behaviours:** Data mining automates the process of finding predictive information in large databases. Questions that traditionally required extensive hands-on analysis can now be answered directly from the data — quickly. A typical example of a predictive problem is targeted marketing. Data mining uses data on past promotional mailings to identify the targets most likely to maximize return on investment in future mailings. Other predictive problems include forecasting bankruptcy and other forms of default, and identifying segments of a population likely to respond similarly to given events.
- **Automated discovery of previously unknown patterns:** Data mining tools sweep through databases and identify previously hidden patterns in one step. An example of pattern discovery is the analysis of retail sales data to identify seemingly unrelated products that are often purchased together. Other pattern discovery problems include detecting fraudulent credit card transactions and identifying anomalous data that could represent data entry keying errors.

Data mining techniques can yield the benefits of automation on existing software and hardware platforms, and can be implemented on new systems as existing platforms are upgraded and new products developed. When data mining tools are implemented on high performance parallel processing systems, they can analyze massive databases in minutes. Faster processing means that users can automatically experiment with more models to understand complex data. High speed makes it practical for users to analyze huge quantities of data. Larger databases, in turn, yield improved predictions.

Databases can be larger in both depth and breadth:

- **More columns:** Analysts must often limit the number of variables they examine when doing hands-on analysis due to time constraints. Yet variables that are discarded because they seem unimportant may carry information about unknown patterns. High performance data mining allows users to explore the full depth of a database, without preselecting a subset of variables.
- **More rows:** Larger samples yield lower estimation errors and variance, and allow users to make inferences about small but important segments of a population.

The most commonly used techniques in data mining are:

- **Artificial neural networks:** Non-linear predictive models that learn through training and resemble biological neural networks in structure.
- **Decision trees:** Tree-shaped structures that represent sets of decisions. These decisions generate rules for the classification of a dataset. Specific decision tree methods include Classification and Regression Trees (CART) and Chi Square Automatic Interaction Detection (CHAID).
- **Genetic algorithms:** Optimization techniques that use processes such as genetic combination, mutation, and natural selection in a design based on the concepts of evolution.
- **Nearest neighbour method:** A technique that classifies each record in a dataset based on a combination of the classes of the k record(s) most similar to it in a historical dataset. Sometimes called the k-nearest neighbor technique.
- **Rule induction:** The extraction of useful if-then rules from data based on statistical significance.

Many of these technologies have been in use for more than a decade in specialized analysis tools that work with relatively small volumes of data. These capabilities are now evolving to integrate directly with industry-standard data warehouse and OLAP platforms.

1.7 DATA PROCESSING

In data mining process, data are usually "collected" from the existing databases, data warehouses, and data marts. Data preprocessing usually includes at least two common tasks:

1. **Outlier detection (and removal):** Outliers are unusual data values that are not consistent with most observations. Commonly, outliers result from measurement errors, coding and recording errors, and, sometimes, are natural, abnormal values. Such non-representative samples can seriously affect the model produced later. There are two strategies for dealing with outliers:

   (*a*) Detect and eventually remove outliers as a part of the preprocessing phase, or

   (*b*) Develop robust modeling methods that are insensitive to outliers.
2. **Scaling, encoding, and selecting features:** Data preprocessing includes several steps such as variable scaling and different types of encoding. For example, one feature with the range [0, 1] and the other with the range [–100, 1000] will not have the same weights in the applied technique;

they will also influence the final data-mining results differently. Therefore, it is recommended to scale them and bring both features to the same weight for further analysis. Also, application-specific encoding methods usually achieve dimensionality reduction by providing a smaller number of informative features for subsequent data modelling.

These two classes of preprocessing tasks are only illustrative examples of a large spectrum of preprocessing activities in a data-mining process. Data-preprocessing steps should not be considered completely independent from other data-mining phases. In every iteration of the data-mining process, all activities, together, could define new and improved data sets for subsequent iterations.

Generally, a good preprocessing method provides an optimal representation for a data-mining technique by incorporating a priori knowledge in the form of application—specific scaling and encoding.

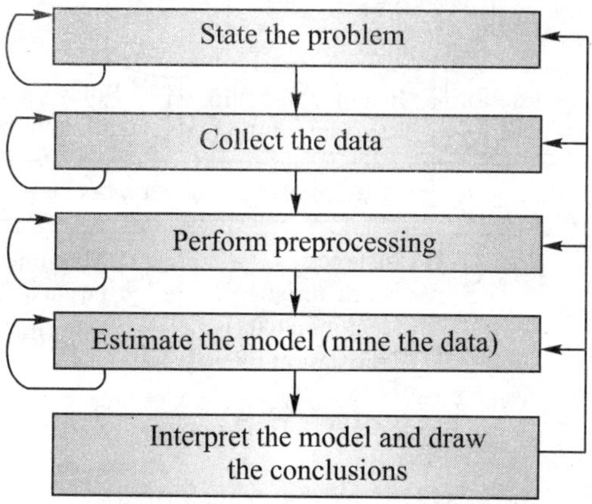

Figure 1.2. Data Mining Process

1.8 DATA CLEANING

Data cleaning, also called *data cleansing* or *scrubbing*, deals with detecting and removing errors and inconsistencies from data in order to improve the quality of data. Data quality problems are present in single data collections, such as files and databases, *e.g.*, due to misspellings during data entry, missing information or other invalid data. When multiple data sources need to be integrated, *e.g.*, in data warehouses, federated database systems or global web-based information systems, the need for data cleaning increases significantly. This is because the sources often contain redundant data in different representations. In order to provide access to accurate and consistent data, consolidation of different data representations and elimination of duplicate information become necessary. Data warehouses require and provide extensive support for data cleaning. They load and continuously refresh huge amounts of data from a variety of sources so the probability that some of the sources contain "dirty data" is high. Furthermore, data warehouses are used for decision making, so that the correctness of their data is vital to avoid wrong conclusions. For instance, duplicated or missing information will produce incorrect or misleading statistics ("garbage in, garbage out"). Due to the wide range of possible data inconsistencies and the sheer data volume, data cleaning is considered to be one of the biggest problems in data warehousing.

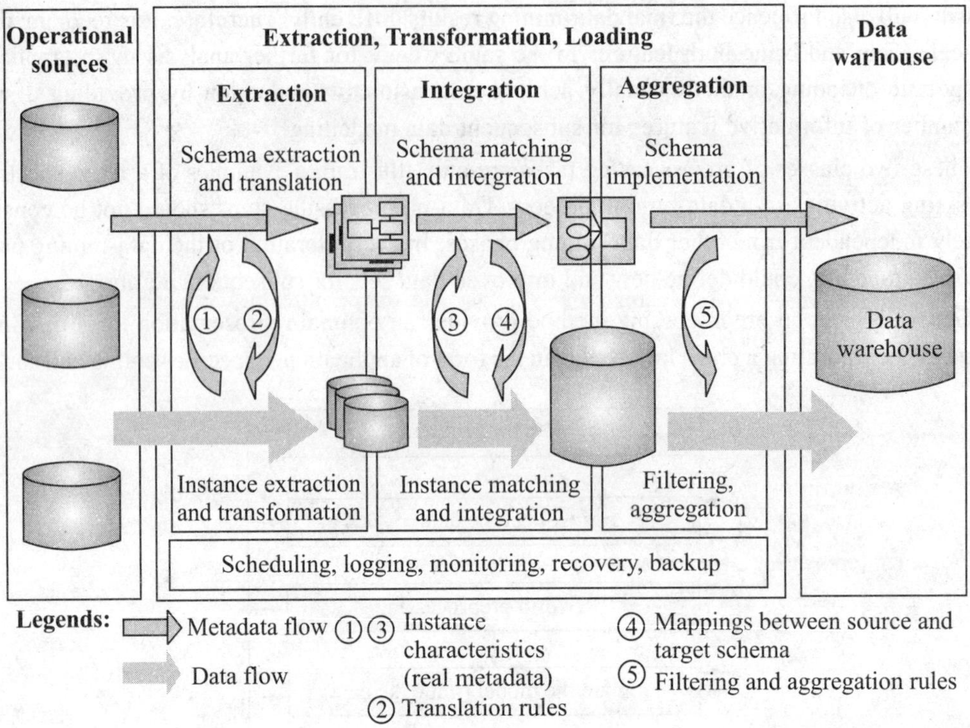

Figure 1.3. ETL Process

During the so-called ETL process (extraction, transformation, loading), illustrated in Fig. 1.3, further data transformations deal with schema/data translation and integration, and with filtering and aggregating data to be stored in the warehouse. As indicated in Fig. 1.3, all data cleaning is typically performed in a separate data staging area before loading the transformed data into the warehouse. A large number of tools of varying functionality is available to support these tasks, but often a significant portion of the cleaning and transformation work has to be done manually or by low-level programs that are difficult to write and maintain.

Federated database systems and web-based information systems face data transformation steps similar to those of data warehouses. In particular, there is typically a *wrapper* per data source for extraction and a *mediator* for integration. So far, these systems provide only limited support for data cleaning, focusing instead on data transformations for schema translation and schema integration. Data is not pre-integrated as for data warehouses but needs to be extracted from multiple sources, transformed and combined during query runtime. The corresponding communication and processing delays can be significant, making it difficult to achieve acceptable response times. The effort needed for data cleaning during extraction and integration will further increase response times but is mandatory to achieve useful query results.

A data cleaning approach should satisfy several requirements. First of all, it should detect and remove all major errors and inconsistencies both in individual data sources and when integrating multiple sources. The approach should be supported by tools to limit manual inspection and programming effort and be extensible to easily cover additional sources. Furthermore, data cleaning should not be performed in isolation but together with schema-related data transformations based on comprehensive metadata. Mapping functions for data cleaning and other data transformations should be specified in a declarative way and be reusable for other data sources as well as for query processing. Especially for data warehouses, a workflow infrastructure should be supported to execute all data transformation steps for multiple sources and large data sets in a reliable and efficient way.

Data Cleaning Approaches

In general, data cleaning involves several phases

(*a*) *Data analysis:* In order to detect which kinds of errors and inconsistencies are to be removed, a detailed data analysis is required. In addition to a manual inspection of the data or data samples, analysis programs should be used to gain metadata about the data properties and detect data quality problems.

(*b*) *Definition of transformation workflow and mapping rules:* Depending on the number of data sources, their degree of heterogeneity and the "dirtyness" of the data, a large number of data transformation and cleaning steps may have to be executed. Sometime, a schema translation is used to map sources to a common data model; for data warehouses, typically a relational representation is used. Early data cleaning steps can correct single-source instance problems and prepare the data for integration. Later steps deal with schema/data integration and cleaning multi-source instance problems, *e.g.,* duplicates. For data warehousing, the control and data flow for these transformation and cleaning steps should be specified within a workflow that defines the ETL process. The schema-related data transformations as well as the cleaning steps should be specified by a declarative query and mapping language as far as possible, to enable automatic generation of the transformation code. In addition, it should be possible to invoke user-written cleaning code and special purpose tools during a data transformation workflow. The transformation steps may request user feedback on data instances for which they have no built-in cleaning logic.

(*c*) *Verification:* The correctness and effectiveness of a transformation workflow and the transformation definitions should be tested and evaluated, *e.g.,* on a sample or copy of the source data, to improve the definitions if necessary. Multiple iterations of the analysis, design and verification steps may be needed, *e.g.,* since some errors only become apparent after applying some transformations.

(*d*) *Transformation:* Execution of the transformation steps either by running the ETL workflow for loading and refreshing a data warehouse or during answering queries on multiple sources.

(*e*) *Backflow of cleaned data:* After (single-source) errors are removed, the cleaned data should also replace the dirty data in the original sources in order to give legacy applications the improved data too and to avoid redoing the cleaning work for future data extractions. For data warehousing, the cleaned data is available from the data staging area.

1.9 DATA INTEGRATION AND TRANSFORMATION

It is likely that your data analysis task will involve data integration, which combines data from multiple sources into a coherent data store, as in data warehousing. These sources may include multiple databases, data cubes, or flat files.

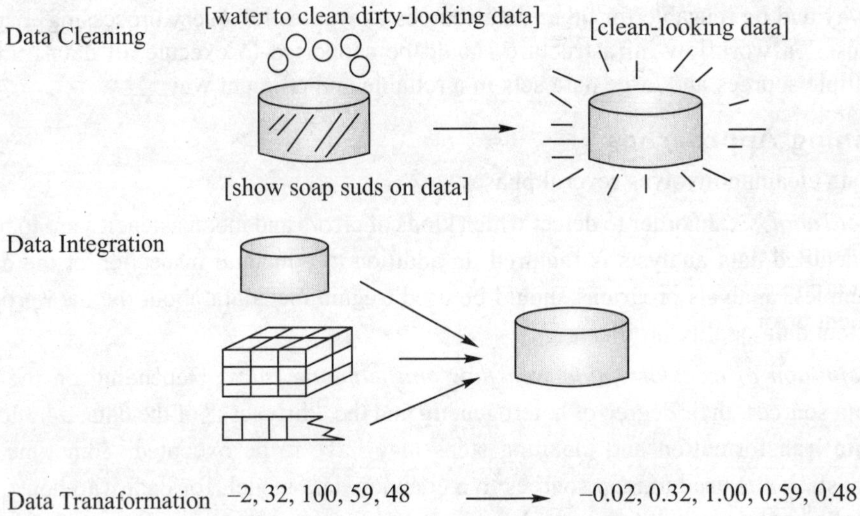

Figure 1.4. Data Pre-processing

There are a number of issues to consider during data integration. Schema integration can be tricky. How can like real-world entities from multiple data sources be "matched up"? This is referred to as the entity identification problem. For example, how can the data analyst or the computer be sure that customer id in one database, and customer number in another refer to the same entity? Databases and data warehouses typically have metadata–that is, data about the data. Such metadata can be used to help avoid errors in schema integration.

Redundancy is another important issue. An attribute may be redundant if it can be derived from another table, such as annual revenue. Inconsistencies in attribute or dimension naming can also cause redundancies in the resulting data set.

Some redundancies can be detected by correlation analysis. For example, given two attributes, such analysis can measure how strongly one attribute implies the other, based on the available data.

In data transformation, the data are transformed or consolidated into forms appropriate for mining. Data transformation can involve the following:

1. Smoothing, which works to remove the noise from data. Such techniques include binning, clustering, and regression.
2. Aggregation, where summary or aggregation operations are applied to the data. For example, the daily sales data may be aggregated so as to compute monthly and annual total amounts. This step is typically used in constructing a data cube for analysis of the data at multiple granularities.
3. Generalization of the data, where low level or "primitive" (raw) data are replaced by higher level concepts through the use of concept hierarchies. For example, categorical attributes, like street, can be generalized to higher level concepts, like city or county. Similarly, values for numeric attributes, like age, may be mapped to higher level concepts, like young, middle-aged, and senior.
4. Normalization, where the attribute data are scaled so as to fall within a small specified range, such as -1.0 to 1.0, or 0 to 1.0.
5. Attribute construction (or feature construction), where new attributes are constructed and added from the given set of attributes to help the mining process.

1.10 DATA REDUCTION

Data Reduction is the transformation of numerical or alphabetical digital information derived empirical or experimentally into a corrected, ordered, and simplified form.

When the information is derived from instrument readings then there may also be a transformation from analog to digital form. When the data are already in digital form the 'reduction' of the data typically involves some editing, scaling, coding, sorting, collating, and producing tabular summaries, and when the observations are discrete but the underlying phenomenon is continuous then smoothing and interpolation are likely to be needed. Often the data reduction is undertaken in the presence of reading or measurement errors. Some idea of the nature of these errors is needed before the most likely value may be determined.

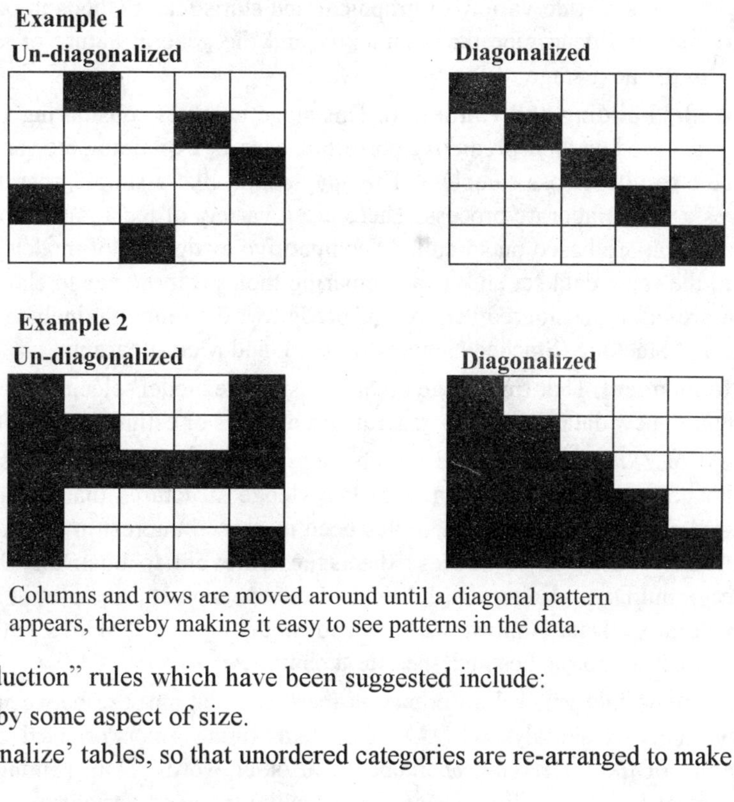

Columns and rows are moved around until a diagonal pattern appears, thereby making it easy to see patterns in the data.

"Data Reduction" rules which have been suggested include:

1. Order by some aspect of size.
2. 'Diagonalize' tables, so that unordered categories are re-arranged to make patterns easier to see.
3. Use averages to provide a visual focus as well as a summary.
4. Use layout and labeling to guide the eye.
5. Remove Chartjunk, such as pictures and lines.
6. Give a brief verbal summary.

1.11 DATA MINING STATISTICS

Data Mining is an analytic process designed to explore data in search of consistent patterns or systematic relationships between variables, and then to validate the findings by applying the detected patterns to new subsets of data. The ultimate goal of data mining is prediction—and predictive data mining is the most common type of data mining and one that has the most direct business applications. The process of data mining consists of three stages:

(*a*) the initial exploration,

(*b*) model building or pattern identification with validation/verification

(c) deployment (*i.e.,* the application of the model to new data in order to generate predictions).

Stage 1: Exploration. This stage usually starts with data preparation which may involve cleaning data, data transformations, selecting subsets of records and—in case of data sets with large numbers of variables—performing some preliminary feature selection operations to bring the number of variables to a manageable range depending on the statistical methods which are being considered. Then, depending on the nature of the analytic problem, this first stage of the process of data mining may involve anywhere between a simple choice of straightforward predictors for a regression model, to elaborate exploratory analyses using a wide variety of graphical and statistical methods in order to identify the most relevant variables and determine the complexity and the general nature of models that can be taken into account in the next stage.

Stage 2: Model building and validation. This stage involves considering various models and choosing the best one based on their predictive performance (*i.e.*, explaining the variability in question and producing stable results across samples). This may sound like a simple operation, but in fact, it sometimes involves a very elaborate process. There are a variety of techniques developed to achieve that goal–many of which are based on so-called "competitive evaluation of models," that is, applying different models to the same data set and then comparing their performance to choose the best. These techniques–which are often considered the core of predictive data mining–include: Bagging (Voting, Averaging), Boosting, Stacking (Stacked Generalizations), and Meta-Learning.

Stage 3: Deployment. That final stage involves using the model selected as best in the previous stage and applying it to new data in order to generate predictions or estimates of the expected outcome.

The concept of *Data Mining* is becoming increasingly popular as a business information management tool where it is expected to reveal knowledge structures that can guide decisions in conditions of limited certainty. Recently, there has been increased interest in developing new analytic techniques specifically designed to address the issues relevant to business *Data Mining* (*e.g.*, Classification Trees), but Data Mining is still based on the conceptual principles of statistics including the traditional Exploratory Data Analysis (EDA) and modeling and it shares with them both some components of its general approaches and specific techniques.

However, an important general difference in the focus and purpose between Data Mining and the traditional Exploratory Data Analysis (EDA) is that Data Mining is more oriented towards applications than the basic nature of the underlying phenomena. In other words, Data Mining is relatively less concerned with identifying the specific relations between the involved variables.

Following are the statistical fields where data mining finds its applications:

Decision Trees: The Decision Tree is one of the most popular classification algorithms in current use in Data Mining and Machine Learning.

Information Gain: This step through the ideas from Information Theory that eventually lead to Information Gain—one of the most popular measures of association currently used in data mining.

Probability for Data Miners: It is a useful investment to be completely happy with probability before venturing into advanced algorithms from data mining, machine learning or applied statistics.

Probability Density Functions: It deals with real-valued random variables, probability density functions, and how to deal with multivariate (*i.e.* high dimensional) probablity densities.

Gaussians: Gaussians, both the friendly univariate kind, and the slightly-reticent-but-nice-when-you-get-to-know-them multivariate kind are extremely useful in many parts of statistical data mining, including many data mining models in which the underlying data assumption is highly non-Gaussian.

Maximum Likelihood Estimation: MLE is a solid tool for learning parameters of a data mining model. It is a methodlogy which tries to do two things. First, it is a reasonably well-principled way to work out what computation you should be doing when you want to learn some kinds of model from data. Second, it is often fairly computationally tractable.

Cross-Validation: Cross-validation is one of several approaches to estimating how well the model you've just learned from some training data is going to perform on future as-yet-unseen data.

Instance-based learning (aka Case-based or Memory-based or non-parametric): Over a century old, this form of data mining is still being used very intensively by statisticians and machine learners alike.

Eight Regression Algorithms: The regression algorithms include Regression Trees, Cascade Correlation, Group Method Data Handling (GMDH), Multivariate Adaptive Regression Splines (MARS), Multilinear Interpolation, Radial Basis Functions, Robust Regression, Cascade Correlation + Projection Pursuit

Predicting Real-valued Outputs: An introduction to regression: Various topics include in this are Varying noise, Non-linear regression (very briefly), Polynomial Regression, Radial Basis Functions, Robust Regression, Regression Trees, Multilinear Interpolation and MARS.

Learning Bayesian Networks: This is an area of active research by many research group.

Gaussian Mixture Models: Gaussian Mixture Models (GMMs) are among the most statistically mature methods for clustering (though they are also used intensively for density estimation).

K-means and Hierarchical Clustering: K-means is the most famous clustering algorithm.

Hidden Markov Models: This simulates a very common phenomenon- there is some underlying dynamic system running along according to simple and uncertain dynamics, but we can't see it. All we can see are some noisy signals arising from the underlying system. From those noisy observations we want to do things like predict the most likely underlying system state, or the time history of states, or the likelihood of the next observation. This has applications in fault diagnosis, robot localization, computational biology, speech understanding and many other areas.

PAC Learning: PAC stands for "Probably Approximately Correct" and concerns a nice formalism for deciding how much data you need to collect in order for a given classifier to achieve a given probability of correct predictions on a given fraction of future test data. The resulting estimate is somewhat conservative but still represents an interesting avenue by which computer science has tried to muscle in on the kind of analytical problem that you would normally find in a statistics department.

Game Tree Search Algorithms, including Alpha-Beta Search: Introduction to algorithms for computer game playing.

Hill Climbing, Simulated Annealing and Genetic Algorithms: Some very useful algorithms, to be used only in case of emergency.

1.12 GUIDELINES FOR SUCCESSFUL DATA MINING

As data mining moves into the mainstream, more classically schooled statisticians and business analysts are exploring ways in which data mining can help them extract new information from their vast corporate databases. Since data mining differs from traditional statistics, following are the few tips for successful data mining:

(*a*) Mine significantly more data.

(*b*) Create new variables to tease more information out of your data.

(*c*) Take a shallow dive into data first.

(*d*) Rapidly build many exploratory predictive models.

(*e*) Forget about traditional data hygiene practices.

(*f*) Enrich the data with external data.

(*g*) Cluster your customers first, and then build multiple targeted predictive models.

(*h*) Automated model building.

(*i*) Feed the models a better "Balanced Fuel Mixture" of data.

(*j*) Demystify neural networks and clusters by reverse engineering them using CRT models.

(*k*) Use predictive modeling to impute missing values.

(*l*) Build multiple models and form "Panel of experts" predictive models.

1.13 ISSUES IN DATA MINING

Data mining algorithms embody techniques that have sometimes existed for many years, but have only lately been applied as reliable and scalable tools that time and again outperform older classical statistical methods. While data mining is still in its infancy, it is becoming a trend and ubiquitous. Before data mining develops into a conventional, mature and trusted discipline, many still pending issues have to be addressed. Some of these issues are addressed below. Note that these issues are not exclusive and are not ordered in any way.

Security and social issues: Security is an important issue with any data collection that is shared and/or is intended to be used for strategic decision-making. In addition, when data is collected for customer profiling, user behavior understanding, correlating personal data with other information, etc., large amounts of sensitive and private information about individuals or companies is gathered and stored. This becomes controversial given the confidential nature of some of this data and the potential illegal access to the information. Moreover, data mining could disclose new implicit knowledge about individuals or groups that could be against privacy policies, especially if there is potential dissemination of discovered information. Another issue that arises from this concern is the appropriate use of data mining. Due to the value of data, databases of all sorts of content are regularly sold, and because of the competitive advantage that can be attained from implicit knowledge discovered, some important information could be withheld, while other information could be widely distributed and used without control.

User interface issues: The knowledge discovered by data mining tools is useful as long as it is interesting, and above all understandable by the user. Good data visualization eases the interpretation of data mining results, as well as helps users better understand their needs. Many data exploratory analysis tasks are significantly facilitated by the ability to see data in an appropriate visual presentation. There are many visualization ideas and proposals for effective data graphical presentation. However, there is still much research to accomplish in order to obtain good visualization tools for large datasets that could be used to display and manipulate mined knowledge. The major issues related to user interfaces and visualization are "screen real-estate", information rendering, and interaction. Interactivity with the data and data mining results is crucial since it provides means for the user to focus and refine the mining tasks, as well as to picture the discovered knowledge from different angles and at different conceptual levels.

Mining methodology issues: These issues pertain to the data mining approaches applied and their limitations. Topics such as versatility of the mining approaches, the diversity of data available, the dimensionality of the domain, the broad analysis needs (when known), the assessment of the knowledge discovered, the exploitation of background knowledge and metadata, the control and handling of noise in data, etc. are all examples that can dictate mining methodology choices. For instance, it is often desirable to have different data mining methods available since different approaches may perform differently depending upon the data at hand. Moreover, different approaches may suit and solve user's needs differently.

Most algorithms assume the data to be noise-free. This is of course a strong assumption. Most datasets contain exceptions, invalid or incomplete information, etc., which may complicate, if not obscure, the analysis process and in many cases compromise the accuracy of the results. As a consequence, data preprocessing (data cleaning and transformation) becomes vital. It is often seen as lost time, but data cleaning, as time-consuming and frustrating as it may be, is one of the most important phases in the knowledge discovery process. Data mining techniques should be able to handle noise in data or incomplete information.

More than the size of data, the size of the search space is even more decisive for data mining techniques. The size of the search space is often depending upon the number of dimensions in the domain space. The search space usually grows exponentially when the number of dimensions increases. This is known as the *curse of dimensionality*. This "curse" affects so badly the performance of some data mining approaches that it is becoming one of the most urgent issues to solve.

Performance issues: Many artificial intelligence and statistical methods exist for data analysis and interpretation. However, these methods were often not designed for the very large data sets data mining is dealing with today. Terabyte sizes are common. This raises the issues of scalability and efficiency of the data mining methods when processing considerably large data. Algorithms with exponential and even medium-order polynomial complexity cannot be of practical use for data mining. Linear algorithms are usually the norm. In same theme, sampling can be used for mining instead of the whole dataset. However, concerns such as completeness and choice of samples may arise. Other topics in the issue of performance are *incremental updating*, and parallel programming. There is no doubt that parallelism can help solve the size problem if the dataset can be subdivided and the results can be merged later. Incremental updating is important for merging results from parallel mining, or updating data mining results when new data becomes available without having to re-analyze the complete dataset.

Data source issues: There are many issues related to the data sources, some are practical such as the diversity of data types, while others are philosophical like the data glut problem. We certainly have an excess of data since we already have more data than we can handle and we are still collecting data at an even higher rate. If the spread of database management systems has helped increase the gathering of information, the advent of data mining is certainly encouraging more data harvesting. The current practice is to collect as much data as possible now and process it, or try to process it, later. The concern is whether we are collecting the right data at the appropriate amount, whether we know what we want to do with it, and whether we distinguish between what data is important and what data is insignificant. Regarding the practical issues related to data sources, there is the subject of heterogeneous databases and the focus on diverse complex data types. We are storing different types of data in a variety of repositories. It is difficult to expect a data mining system to effectively and efficiently achieve good mining results on all kinds of data and sources. Different kinds of data and sources may require distinct algorithms and methodologies. Currently, there is a focus on relational databases and data warehouses, but other approaches need to be pioneered for other specific complex data types. A versatile data mining tool, for all sorts of data, may not be realistic. Moreover, the proliferation of heterogeneous data sources, at structural and semantic levels, poses important challenges not only to the database community but also to the data mining community.

SUMMARY

- Data mining software analyzes relationships and patterns in stored transaction data based on open-ended user queries.
- CART and CHAID are decision tree techniques used for classification of a dataset.
- *Data cleaning* and *data integration* can be performed together as a pre-processing phase to generate a data warehouse.
- *Data selection* and *data transformation* can also be combined where the consolidation of the data is the result of the selection.
- Data characterization is a summarization of general features of objects in a target class, and produces what is called *characteristic rules*.
- Classification approaches normally use a *training set* where all objects are already associated with known class labels.
- The classification algorithm learns from the training set and builds a model.
- Evolution analysis models evolutionary trends in data, which consent to characterizing, comparing, classifying or clustering of time related data.
- Deviation analysis considers differences between measured values and expected values, and attempts to find the cause of the deviations from the anticipated values.
- Data mining is being put into use and studied for databases, including relational databases, object-relational databases and object-oriented databases, data warehouses, transactional databases, unstructured and semi-structured repositories such as the World Wide Web, advanced databases such as spatial databases, multimedia databases, time-series databases and textual databases, and even flat files.
- Data cleaning should not be performed in isolation but together with schema-related data transformations based on comprehensive metadata.
- Mapping functions for data cleaning and other data transformations should be specified in a declarative way and be reusable for other data sources as well as for query processing.
- In data transformation, the data are transformed or consolidated into forms appropriate for mining.
- Data Reduction is the transformation of numerical or alphabetical digital information derived empirical or experimentally into a corrected, ordered, and simplified form.

REVIEW QUESTIONS

1. What do you understand by data mining? How data mining works?
2. Describe the classification of data mining systems.
3. List out data mining issues and operations.
4. What is knowledge discovery process? How is it different from data mining process?
5. What are the different data mining techniques?
6. Write short notes on:
   (*a*) Data Cleaning
   (*b*) Data Processing
   (*c*) Data Integration and Transformation
   (*d*) Data Reduction.

■■■

CHAPTER 2

Data Preprocessing, Language and Concept Description

Data mining can be classified into two categories; one is descriptive data mining and the other is predictive data mining. While descriptive data mining describes the data set in a concise and summative manner and presents interesting properties of the data; Predictive data mining analyzes the data in order to construct one or a set of models, and attempts to predict the behavior of new data sets. Data bases usually store large amounts of data in great detail. Users often like to view sets of summarized data in concise, descriptive terms. Such data descriptions provide an overall picture of a class of data or distinguish it from a set of comparative classes. Moreover, users like the simplicity and flexibility of having data sets described at different levels of granularity and from different angles. Such descriptive data mining is called *concept description* and forms an important component of data mining.

In the previous chapter we studied about data mining. In this chapter we will see data preprocessing. The databases store different types of data. Now the database may contain noisy data, inconsistent data and missing data. This is because the database may contain huge amount of data (in megabytes or gigabytes). The data stored in a database may be extracted from multiple distinct data sources. The data is cleaned, integrated, reduced, transformed and stored in a database. Now if the data is of low data quality then it will result in low quality data mining.

We have seen data cleaning, data integration, data transformation and data reduction in brief. In this chapter we will study these concepts in detail. We will also study the role played by data preprocessing in data mining. The data preprocessing can be compared to cooking of food. When a mixed vegetable is cooked then we will use various vegetables (potato, carrot, peas, etc). We will first wash the vegetables and utensils. Then we will remove the peel of the vegetables. We may also want to add salt, pepper according to taste while it is being cooked. Now let us understand the comparison. Washing and removing peel is similar to data reduction. Adding of vegetables, salt, pepper and turmeric is similar to data integration. Cooking is comparable to data transformation because cooking involves a chemical reaction that transforms raw vegetables to cooked vegetables. However the concepts of data integration, cleaning and transformation are entirely different. We just want to compare so that to make the understanding simple. We will study these concepts in the later sections.

The simplest kind of descriptive data mining is concept description. A concept refers to a collection of data such as number_of_customers, engineering_students, and so on. As a data mining task, concept description is not a simple enumeration of the data. Instead, concept description generates descriptions for characterization and comparison of the data. It is sometimes called class description, when the concept to be described refers to a class of objects. Characterization provides a concise and crisp summarization of the given collection of the data, while concept or class comparison (also known as discrimination) provides discriminations comparing two or more collections of data. Concept description has close ties with the data generalization.

2.1 PREPROCESSING

Data quality is an important concept in data mining. The data quality can be measured in accuracy, completeness, consistency, timeliness and interpretability. The data in the database lacks quality if it contains noisy data, inconsistent data and missing data. There may be many reasons which may cause the data to be faulty like the data source from where the data is collected may be faulty. There may be human or computer errors like when data values are entered in the database. There may be data transmission errors which may be result of limited buffer size. One other reason for incorrect data may be if there is no standard convention for entering data.

Other type of error is incomplete data. Sometimes some attribute values are not recorded. This may be if the attribute value is not relevant. But if the attribute value is relevant then incomplete data may result if the value of that attribute is not available or if there is a malfunction in the equipment.

Timeliness is an important factor in data quality. It occurs when the required data is added at a later point of time. So some decisions might have been taken before the whole data is available. Believability means if the data should be trusted or not. Interpretability means if it is easy to understand the data or the data can be interpreted or not. If the data can be analyzed and interpreted very easily then the data is of good quality otherwise it is not.

The major tasks involved in data preprocessing are:

1. Data Cleaning
2. Data Integration
3. Data Reduction
4. Data Transformation

Data Cleaning means to clean the data. The data is cleaned by filling in missing values, smoothing noisy data, removing outliers and inconsistencies. The data cleaning is necessary because if the data is not cleaned then the mining results resulting may be faulty and cannot be trusted. The dirty data can cause unreliable output. Hence data cleaning is required. For details, study chapter 1.

Data Integration means integrating data from many sources(databases, data cubes or files). The data integration is required because we need to combine data from different data sources. In databases the names of attributes of same data may be different in different databases. For example an employee id may be referred as emp_id or e_id. There are so many redundancies and inconsistencies in the database which hampers knowledge discovery process. In data warehouse cleaning and integration acts as a major step. We will study data warehousing in the later chapters. Data redundancies occurring from data integration can again be removed by data cleaning. In data transformation, the data are transformed or consolidated into forms appropriate for mining. Refer chapter 1 section 9: Data Integration and Transformation.

Data reduction is the step needed when the amount of data selected is very large. Now if we use this huge amount of data then it will slow the mining process. So we require the reduced set of data that can be successfully and speedily employed for the mining process. The reduction gives a set of data which is very small in volume. Data reduction are of two types dimensionality reduction and numerosity reduction. We will study dimensionality reduction and numerosity reduction in this chapter in the last section data reduction.

2.2 DISCRETIZATION

Discretization is a process in which continuous or real valued attributes can be discretized. By discretization we mean that the values of an attribute are partitioned into intervals. These intervals are the discrete values of the continuous range of values of attributes. The discrete values can be compared to the grading system of the university. If you score more than 80 percent you get an A grade. If your percentage is between 80 and 70 then your grade is B and so on. We will also see the same with the help of a diagram. The figure 2.1 illustrates discretization for the attribute marks. The bottom of this figure shows an ordered list of continuous marks. The continuous marks are discretized into intervals. The grades are shown at the top of the interval in the figure.

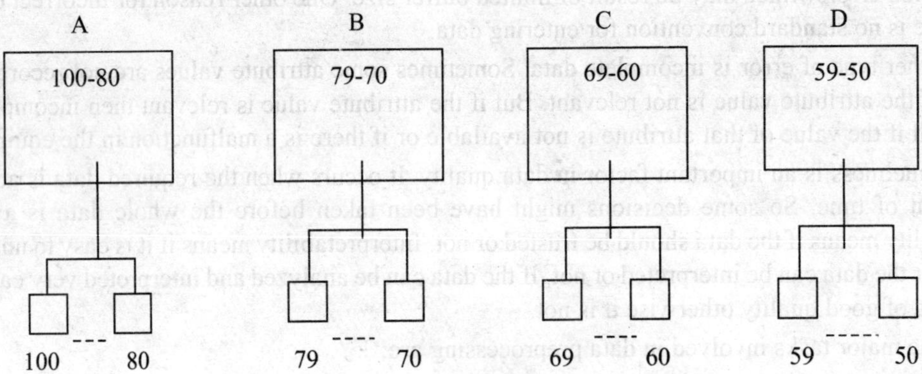

Figure 2.1. Discretization of Attribute Marks

There are various types of discretiztion methods. One of the simplest methods is the equal width method. In this method the continuous range of an attribute is divided into equal width intervals like 0-2,2-4, 4-6, 6-8, 8-10.

The discretization methods can be divided into types: Class blind methods and class driven methods. Class blind methods divide the attribute into discrete values without considering classes or goal attribute values like in equal width method discretiztion is done without goal attribute values.

Class driven methods discretize the attribute values considering the class. The class driven methods are more suitable if the resulting interval is to be used in classification algorithm. For example, if your percentage lies between 40 and 100 then you are passed but if your percentage lies between 0 and 40 you are fail. Whereas the class blind method like equal width method divides the attribute values into equal intervals say 0-20, 20-40 and so on. In heuristic algorithms class driven methods are used as iterative algorithms. In each iteration, discretization is done and particular interval that satisfies some condition is selected. This discretization is performed until some stopping condition is reached.

The advantages of discretization are as follows:

1. Discretization is a data abstraction process that transforms lower level data into higher level data and hence improves the comprehensiveness of the discovery rules.
2. Discretization reduces the time taken in running a classification and decision tree algorithm.
3. Some rule discovery algorithms cannot be used with attributes with continuous values. Hence some discretization of these values is required.

The disadvantages of discretization are as follows:

1. The discretization transforms the data from one level to the other. In this way the accuracy of discovered knowledge may be reduced.
2. Discretization may increase or decrease the accuracy of the classification. It varies on original data and data mining algorithm adopted.

2.3 CONCEPT HIERARCHY GENERATION

Before studying about concept hierarchy generation we should understand the meaning of nominal data attributes. Nominal attributes are those attributes that have finite number of distinct values for example department. Defining concept hierarchy is a difficult job. Mostly the hierarchies are defined at the schema definition level. Concept hierarchy is used to transform the data into multiple level of granularity. There are four methods for generation of concept hierarchies for nominal data.

1. **Partial ordering of attributes:** Concept hierarchies involve a group of attributes. Concept hierarchy can be defined by specifying partial or total ordering of attributes. Location dimension of a warehouse can be defined as city < state < country < continent.
2. **Specifying portion of hierarchy:** In large database it would be impossible to define concept hierarchy by partial ordering of attributes. Hence portion of the hierarchy can be explicitly grouped.
3. **Specifying set of attributes:** The concept hierarchy can be defined by user by specifying set of attributes without partial grouping of attributes.
4. **Specifying partial set of attributes:** Sometimes the user may not have the clear idea of what should be included in the hierarchy hence partial set of hierarchy attributes may be included. For example user might have included only city and state.

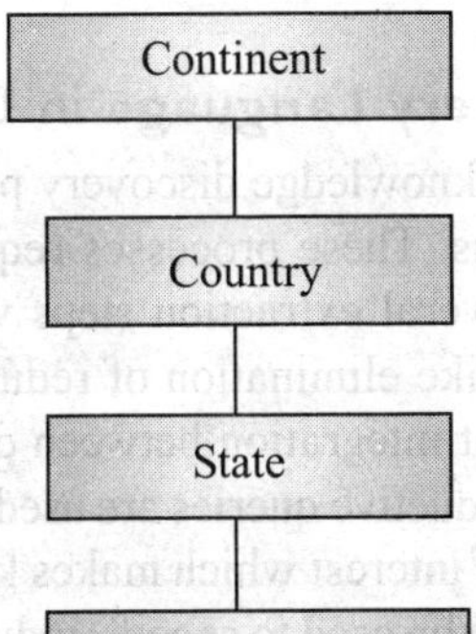

Figure 2.2. Concept hierarchy in dimension location

2.4 DATA MINING PRIMITIVES

In the past the data mining tools were implemented on small databases and flat files. But now due to shift to larger databases the data mining algorithms are implemented on larger databases too. The requirements of data mining algorithms are as follows:

1. Efficient sampling methods

2. Storing immediate results while executing queries
3. Capacity to handle large size data
4. Geometric Indexes

Data mining uses Client Server technology. The server technology is essential when it is required to work with large databases.

Data mining routines are of four types:

1. **Pre-processing:** Here the data mining routines take samples, organize attribute values into clusters and compress code tables into appropriate input formats for algorithms. Here historical files are transformed into time series, multi valued attributes are transformed into binary
2. **Nearest neighbour search:** It is interpreted as a generalization of exact identification of records by traditional keys.
3. **N table scan algorithms:** In these algorithms tables are accessed several times during execution.
4. **Zoom scan algorithms:** These algorithms also access tables several times. The use of relational technology improves the speed of execution. These are called zoom algorithms because they zoom into interesting sub parts of database during execution.

2.5 QUERY LANGUAGE

Data Mining algorithms are used to extract different types of patterns from data. We require integrated systems which can deal with patterns and data to discover knowledge which requires inductive database approach. Hence query language has to be designed to support knowledge discovery processes. In the field of association rule mining, query language is required to support the declarative specification of data and pattern manipulations.

2.5.1 Why do we Require Query Language in Data Mining

There are many techniques that support knowledge discovery processes in database which is based on some extracted and materialized patterns. These processes require pre processing manipulations like cleaning the data. They also require several extraction steps with different parameters and patterns. Several post processing manipulations like elimination of redundancies in extracted patterns are also required. In an integrated database (tight integration between data and patterns), ordinary queries are used to access and manipulate data and inductive queries are used to mine, manipulate and apply patterns. Analysts can specify data and patterns of interest which makes knowledge discovery extended querying process. Data mining query language can be used to specify inductive queries on some pattern domains. Data mining query language should support the selection of pre processed data that is to be mined, specification of the kind of patterns to be mined and needed background knowledge, constraints on desired patterns and post process extracted patterns. It should support the closure property i.e. result of a query can be queried. The language should be well defined. But evaluation of data mining queries is very expensive which requires efficient constraint based data mining algorithms. At present there are only few languages in the field of rule association mining.

Some proposals for association rule mining are as follows:

1. **MSQL:** MSQL was designed by Imielinski and Virmani in 1999 at Rutgers University. It uses rules based on descriptors. In a descriptor LHS is an attribute and RHS is a range of values in the domain of LHS. MSQL uses propositional rules of the type A=>B. Here A is conjuncset and B is a descriptor.

2. **MINE RULE:** MINE RULE was designed by Meo et al. in 1998 at the university of Torino and Politecnico di Milano. It is extended SQL coupled with RDBMS where data can be selected using full power of SQL. In MINE RULE an association rule is materialized into relational tables. The extracted form of rule, cardinality of body and the head of desired rules and attributes is decided by the user.
3. **DMQL:** It was designed by Han et al. in 1996 at Simon Fraser University at Canada. It supports various rule extractions like classification rules, comparison rules and association rules. An association rule relates values of two sets of predicates evaluated on the relations of a database. Predicates are of the form P(A, b), where P is a predicate, X is a variable and c is a value in the domain of the attribute.

2.6 GRAPHICAL USER INTERFACES AND ARCHITECTURES

Data mining is an exploratory process. An easy user interface is essential so that even non technical user can also use the data mining tool effectively. The graphic user interface should be of high quality and should be highly interactive because otherwise it would be much difficult for business users to understand it properly. Hence the GUI should be user friendly. Other user interfaces are constructed on top of SQL that serves as query language. But data mining interface does not share any data mining query language. Due to this data mining products are not standardized because there is no any common query language.

Let us generalize the functions of the data mining user interface.

1. The mining interface should be GUI.
2. There should be simple options for business or unskilled users but advanced options for skilled users.
3. Ability to build models in batch mode.
4. Additional applications should be supported by an API (Application Programming Interface).
5. The interface should also include help system.
6. The interface should provide default values to support lower limit records required to support a split in decision tree or number of nodes in hidden layer of NN so that probability for success is maximum for casual users.
7. Interface should provide management of data mining algorithms.

Data Mining Architecture

The components of a data warehouse of a mining system are as follows:

1. **Data Warehouse:** Data warehouse is an information repository where data is stored. It is analogous to the database of a database management system. Data cleaning and integration techniques are performed on data warehouse.
2. **Data Warehouse Server:** Data warehouse server maps the user's data mining request to the relevant data in data warehouse.
3. **Knowledge Base:** Knowledge base acts as a guide to search and evaluate the resulting patterns. It contains concept hierarchies that organize values of attributes into levels of abstraction.
4. **Data Mining Engine:** It is an important component of mining system and is used for characterization, association, classification, cluster analysis and evolution analysis.
5. **Pattern Evolution Module:** This component interacts with the data mining modules to focus on the search of interesting patterns for efficient data mining.

6. **GUI:** This component is responsible for communication between user and the mining system. The user interacts with the system by entering some query. Through the graphic user interface, user can browse the data in the data warehouse, evaluate mined patterns and analyze the different forms of the patterns. A good GUI is very simple and easy and contains help for efficient interaction between users and mining system.

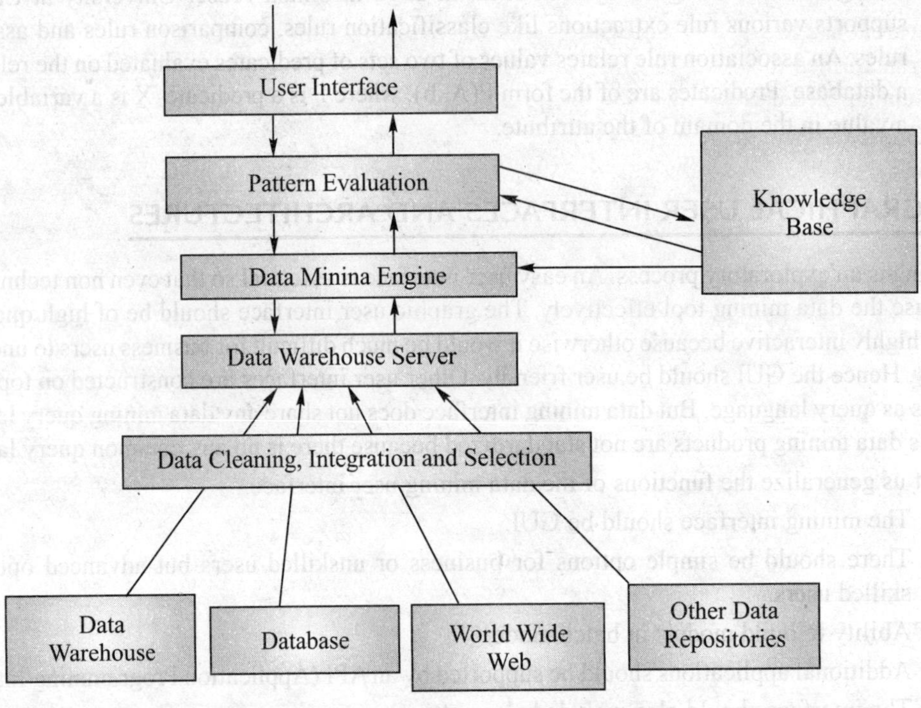

Figure 2.3. Data Mining Architecture

2.7 DATA REDUCTION

Data reduction is the step needed when the amount of data selected is very large. Now if we use this huge amount of data then it will slow the mining process. So we require the reduced set of data that can be successfully and speedily employed for the mining process. The reduction gives a set of data which is very small in volume.

When the information is derived from instrument readings then there may also be a transformation from analog to digital form. When the data are already in digital form the 'reduction' of the data typically involves some editing, scaling, coding, sorting, collating, and producing tabular summaries, and when the observations are discrete but the underlying phenomenon is continuous then smoothing and interpolation are likely to be needed. Often the data reduction is undertaken in the presence of reading or measurement errors. Some idea of the nature of these errors is needed before the most likely value may be determined.

Data reduction techniques obtain a data set that is much smaller in volume but maintains the integrity of the original data. Mining on the smaller data set should be efficient but should produce same results.

1. **Data Cube Aggregation:** Data cube aggregation is an aggregation in which operations are applied to the data in the construction of a data cube. Aggregation is nothing but a technique to merge or consolidate dispersed data into single data. The data that has been summarized is the summarized data.

   Aggregation plays a key role in data warehousing. Aggregation is done to speed up the query processing and improve the overall performance in terms of query response time and data warehouse usability. Aggregation is used when the data extracted from databases has to be transformed to be stored in the data warehouse. Aggregation summarizes different data into a summarized data by removing duplicate data.

   One disadvantage of aggregation is that although the queries are run faster yet some data may be lost.

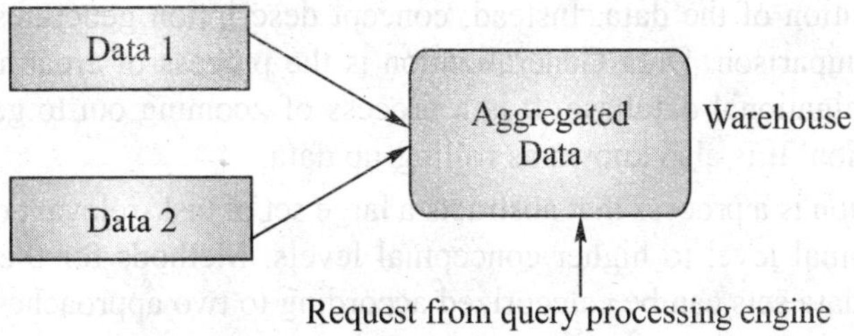

Figure 2.4. Aggregation in warehouses

2. **Dimensionality Reduction:** In dimensionality reduction the number of random variables or attributes under consideration are reduced. It contains wavelet transforms and principal component analysis methods that transform the data into smaller set. It contains a method called attribute subset selection which removes irrelevant and redundant attributes.
3. **Numerosity Reduction:** In these techniques original data is replaced by alternative smaller form of data representation. The techniques may be parametric or non parametric. In parametric methods only data parameters need to be stored and not the actual data. Whereas in non parametric methods reduced representations of data are stored. They include histograms, clustering, sampling and data cube aggregation.
4. **Data Compression:** Data compression techniques obtain a compressed version of the original data. If we can reconstruct the original data from the compressed data without some data being reduced then it is called lossless reduction otherwise if some part of data is lost then it is called lossy reduction. Dimensionality reduction and numerosity reduction techniques are considered as a part of data compression techniques.

2.8 DATA GENERALIZATION

Data and objects in databases often contain detailed information at primitive concept levels. For example, the item relation in sales database may contain attributes describing low-level item information such as item _ID, name, brand, category, supplier, place_made, and price. It is useful to be able to summarize a large set or data and present it at a high conceptual level. For example, summarizing a large set of items relating to Christmas season sales provides a general description of such data, which can be very helpful for sales and marketing managers. This requires an important functionality in data mining: data generalization.

Data generalization summarizes data by replacing relatively low-level values *e.g.*, numeric values for an attribute age, with higher-level concepts *e.g.*, young, middle-aged, and senior, or by reducing the number of dimensions to summarize data in concept space involving fewer dimensions *e.g.*, removing birth date and telephone number when summarizing the behavior of a group of students. Given the large amount of data stored in databases, it is useful to be able to describe concepts in concise terms at generalized (rather than low) levels of abstraction. Allowing datasets to be generalized at multiple levels of abstraction facilitates users in examining the general behavior of the data. Given all Electronics database, for example, instead of examining individual customer transactions, sales managers may prefer to view the data generalized to higher levels, such as summarized by customer groups according to geographic regions, frequency of purchases per group, and customer income. This leads us to the notion of concept description, which is a form of data generalization. A concept typically refers to data collection such as frequent buyers, graduate students, and so on. As a data mining task, concept description is not a simple enumeration of the data. Instead, concept description generates descriptions for data characterization and comparison. Data Generalization is the process of creating successive layers of summary data in an evaluational database. It is a process of zooming out to get a broader view of a problem, trend or situation. It is also known as rolling-up data.

Data generalization is a process that abstracts a large set of task-relevant data in a database from a relatively low conceptual level to higher conceptual levels. Methods for the efficient and flexible generalization of large data sets can be categorized according to two approaches:

(*a*) Data cube (or OLAP) approach

(*b*) Attribute–oriented induction approach

(*a*) **Data cube approach for data generalization:** In the data cube approach (or OLAP approach) to data generalization, the data for analysis are stored in a multidimensional database, or data cube. In general, the data cube approach materializes data cubes by first identifying expensive computations required for frequently-processed queries. These operations typically involve aggregate functions, such as count(), sum(), average(), and max(). The computations are performed, and their results are stored in data cubes. Such computations may be performed for various levels of data abstraction. These materialized views can then be used for decision support, knowledge discovery, and many other applications. A set of attributes may form a hierarchy or a lattice structure, defining a data cube dimension. For example, date may consist of the attributes day, week, month, quarter, and year which form a lattice structure, and a data cube dimension for time. A data cube can store pre-computed aggregate functions for all or some of its dimensions.

The pre-computed aggregates correspond to specified group-by's of different sets or subsets of attributes. Generalization and specialization can be performed on a multidimensional data cube by roll-up or drill-down operations. A roll-up operation reduces the number of dimensions in a data cube, or generalizes attribute values to higher level concepts. A drill-down operation does the reverse. Since many aggregate functions need to be computed repeatedly in data analysis, the storage of pre-computed results in a multidimensional data cube may ensure fast response time and offer flexible views of data from different angles and at different levels of abstraction.

The data cube approach provides an efficient implementation of data generalization, which in turn forms an important function in descriptive data mining. Most commercial data cube implementations confine the data types of dimensions to simple, nonnumeric data and of measures to simple, aggregated numeric values, whereas many applications may require the analysis of more complex data types. Moreover, the data cube approach cannot answer some important questions which concept description can, such as which dimensions should be used in the description, and at what levels should the generalization process reach. Instead, it leaves the responsibility of these decisions to the users.

(*b*) **Attribute-oriented induction:** The Attribute-Oriented Induction (AOI) approach to data generalization and summarization-based characterization was first proposed in 1989, a few years prior to the introduction of the data cube approach. The data cube approach can be considered as a data warehouse-based, pre-computation-oriented, materialized view approach. It performs offline aggregation before an OLAP or data mining query is submitted for processing. On the other hand, the attribute-oriented induction approach, at least in its initial proposal, is a relational database query-oriented, generalization-based, on-line data analysis technique. However, there is no inherent barrier distinguishing the two approaches based on on-line aggregation versus off-line pre-computation. Some aggregations in the data cube can be computed on-line, while off-line pre-computation of multidimensional space can speed up attribute-oriented induction as well. In fact, data mining systems based on attribute-oriented induction, such as DBMiner, have been optimized to include such off-line pre-computation.

The general idea of attribute-oriented induction is to first collect the task-relevant data using a relational database query and then perform generalization based on the examination of the number of distinct values of each attribute in the relevant set of data. The generalization is performed by either attribute removal or attribute generalization (also known as concept hierarchy ascension). Aggregation is performed by merging identical, generalized tuples, and accumulating their respective counts. This reduces the size of the generalized data set. The resulting generalized relation can be mapped into different forms for presentation to the user, such as charts or rules.

2.9 ANALYTICAL CHARACTERIZATION

The first limitation of class characterization for multidimensional data analysis in data warehouses and OLAP tools is the handling of complex objects. The second limitation is the lack of an automated generalization process: the user must explicitly tell the system which dimensions should be included in the class characterization and to how high a level each dimension should be generalized. Actually, each step of generalization or specialization on any dimension must be specified by the user.

Usually, it is not difficult for a user to instruct a data mining system regarding how high a level each dimension should be generalized. For example, users can set attribute generalization thresholds for this, or specify which level a given dimension should reach, such as with the command generalize dimension location to the country level. Even without explicit user instruction, a default value such as 2 to 8 can be set by the data mining system, which allows each dimension to be generalized to a level that contains only 2 to 8 distinct values. If the user is not satisfied with the current level of generalization, he or she can specify dimensions on which drill-down or roll-up operations should be applied.

However, it is nontrivial for users to determine which dimensions should be included in the analysis of class characteristics. Data relations often contain 50 to 100 attributes, and a user may have little knowledge regarding which attributes or dimensions should be selected for effective data mining. A user may include too few attributes in the analysis, causing the resulting mined descriptions to be incomplete or incomprehensive. On the other hand, a user may introduce too many attributes for analysis. Methods should be introduced to perform attribute (or dimension) relevance analysis in order to filter out statistically irrelevant or weakly relevant attributes, and retain or even rank the most relevant attributes for the descriptive mining task at hand. Class characterization which includes the analysis of attribute/dimension relevance is called *analytical characterization.* Class comparison which includes such analysis is called *analytical comparison*. Intuitively, an attribute or dimension is considered highly relevant with respect to a given class if it is likely that the values of the attribute or dimension may be used to distinguish the class from others. For example, it is unlikely that the color of an automobile can be used to distinguish expensive from cheap cars, but the model, make, style, and number of cylinders are likely to be more relevant attributes. Moreover, even within the same dimension, different levels of concepts may have dramatically different powers for distinguishing a class from others. For example, in the birth date dimension, birth day and birth month are unlikely relevant to the salary of employees. However, the birth decade (i.e., age interval) may be highly relevant to the salary of employees. This implies that the analysis of dimension relevance should be performed at multi-levels of abstraction, and only the most relevant levels of a dimension should be included in the analysis.

As already said that attribute/dimension relevance is evaluated based on the ability of the attribute/dimension to distinguish objects of a class from others. When mining a class comparison (or discrimination), the target class and the contrasting classes are explicitly given in the mining query. The relevance analysis should be performed by comparison of these classes. However, when mining class characteristics, there is only one class to be characterized. That is, no contrasting class is specified. It is therefore not obvious what the contrasting class should be for use in the relevance analysis. In this case, typically, the contrasting class is taken to be the set of comparable data in the database which excludes the set of data to be characterized. For example, to characterize graduate students, the contrasting class is composed of the set of students who are registered but are not graduate students.

2.10 ANALYSIS OF ATTRIBUTE RELEVANCE

Intuitively, an attribute or dimension is considered highly relevant with respect to a given class if it is likely that the values of the attribute or dimension may be used to distinguish the class from others. For example, it is unlikely that the color of an automobile can be used to distinguish expensive from cheap cars, but the model, make, style, and number of cylinders are likely to be more relevant attributes. Moreover, even within the same dimension, different levels of concepts may have dramatically different powers for distinguishing a class from others. For example, in the birth date dimension, birth day and birth month are unlikely relevant to the salary of employees. However, the birth decade (i.e., age interval) may be highly relevant to the salary of employees. This implies that the analysis of dimension relevance should be performed at multilevels of abstraction, and only the most relevant levels of a dimension should be included in the analysis.

When mining a class comparison (or discrimination), the target class and the contrasting classes are explicitly given in the mining query. The relevance analysis should be performed by comparison of these classes. However, when mining class characteristics, there is only one class to be characterized. That is, no contrasting class is specified. It is therefore not obvious what the contrasting class should be for use in the relevance analysis. In this case, typically, the contrasting class is taken to be the set of comparable data in the database which excludes the set of the data to be characterized. For example, to characterize graduate students, the contrasting class is composed of the set of students who are registered but are not graduate students.

Analytical comparison performs attribute and dimension relevance analysis in order to filter out irrelevant or weakly relevant attributes to the induction process. Measures of attribute relevance analysis can be used to identify irrelevant or weakly irrelevant attributes that can be excluded from concept description process. The incorporation of pre-processing step into class comparison is referred to as analytical characterization or analytical comparison.

The reason to perform attribute relevance analysis is due to various limitations of class characterization.

- The first limitation to class characterization for multidimensional data analysis is the handling of complex objects.
- The second limitation is the lack of an automated generalized process.

Each step of generalization or specialization on any dimension must be specified by the user. Usually it is not difficult for any user to instruct a data mining system regarding how high a level dimension should be generalized. However, it is non-trivial for the users to determine which dimension should be included in the analysis of class characterization. Data relations often contain 50 to 100 attributes and a user may have a little knowledge regarding which attributes or dimensions should be selected for effective data mining. A user may include too few attributes in the analysis causing the resulting mined descriptions to be incomplete. Also, a user may introduce too many attributes for analysis.

Methods should be introduced to perform attribute relevance analysis on order to filter out statistically irrelevant or weakly relevant attributes and retain the most relevant attributes for the descriptive mining at hand. Class characterization that includes the analysis of dimension relevance is called analytical characterization. Class comparisons that include such analysis is called analytical comparison.

2.11 MINING CLASS COMPARISONS

In many applications, users may not be interested in having a single class described but would prefer to mine a description that compares or distinguishes one class from other comparable classes. Class comparison mines description that distinguishes a target class from its distinguishing classes. The target and contrasting classes must share same attributes and dimensions to be comparable. The steps followed to perform class comparison are as follows:

(*a*) **Data Collection:** The set of relevant data in the database is collected by query processing and is partitioned into target class or set of contrasting classes.

(*b*) **Dimension Relevance Analysis:** If there are many dimensions then dimension relevance analysis should be done to select the highly relevant dimension for further analysis. Correlation based measures can be used for this step.

(*c*) **Synchronous Generalization:** Generalization is performed on the target class to the level controlled by a user or expert specified dimension threshold which results in a prime target class relation. The concept in the contrasting classes are generalized in the same level as those in the prime target class relation forming the prime contrasting class relation.

(*d*) **Presentation of the Derived Comparison:** The resulting class comparison description can be visualized in the form of tables, graphs and rules. This presentation usually includes a contrasting measure such as count percent that reflects the comparison between the target and contrasting classes. The user can adjust the comparison description by applying drill-down, roll-up and other OLAP operations to the contrasting classes, as desired. We will study OLAP operations in the forthcoming chapters.

2.12 STATISTICAL MEASURES IN LARGE DATABASES

Various tasks of data mining such as central tendency and data dispersion, more data characteristics are required to be understood. Measures of central tendency include mean, median, mode, range etc and measures of data dispersion include quartiles, outliers and variance. These descriptive characteristics are of great help in understanding the distribution of data.

(*a*) **Measuring Central Tendency:** It refers to all those methods of statistical analysis by which the average of the statistical series are worked out. Central tendency is measured by calculating arithmetic mean, median, mode and midrange. The purpose of measuring central tendency is brief and simple representation of a group of individual values so that we can quickly grasp the general basis of the units of the group.

(*i*) **Arithmetic mean:** It is the most common and most effective means of calculation of a set of data. Arithmetic mean is the number obtained by adding the values of all the items in the data set and dividing the total by the total number of items in the data set.

(*ii*) **Median:** It is the centrally located value of a set of data such that half of the values of a set of data are above it and half below it. Actually, median is that value of the variable which divides the group into two equal parts, one part comprising all values greater than the median value and the other part comprising all values lower than median value.

(*iii*) **Mode:** the mode for a set of data is the value that occurs most frequently in the data set. Mode is the value which has greatest frequency density in its immediate neighbourhood.

Mode = 3 Median – 2 Mean

(*iv*) **Midrange:** It is the average of largest and smallest value in the data set.

(*b*) **Measuring Dispersion of Data:** The degree to which data tends to spread about an average value is called dispersion of data. Dispersion is the measure of the variation of items. Dispersion of data can be calculated by range, quartile, outliers and boxplots.

(*i*) **Range:** It is the simplest method of measuring dispersion of data. The range of the set of data is the difference between largest and minimum values.

(*ii*) **Quartiles, Outlier, Boxplots:** Quartiles include median that gives indication about the center spread, and shape of a distribution. Boxplots are a way of visualizing a distribution.

(*iii*) **Standard deviation and Variance:** It is the most satisfactory and scientific measure of dispersion. It is also called root mean square deviation as it is the square root of means of the squared deviation from the arithmetic mean.

(*c*) **Graph Displays of Basic Statistical class Description:** Graphic display of statistical class include histograms, quantile plots, *q-q* plots, scatter plots and loess curve. Plotting histograms or frequency histograms is a graphical method summarizing the distribution of a given attribute.

(*i*) **Histogram:** A histogram for an attribute partitions the data distribution of the attribute into disjoint subsets, or buckets. The width of each bucket is uniform.

(*ii*) **Quantile Plot:** It is a simple and effective way to have a first look at a univariate data distribution. It displays all data related to a given attribute and plots quantile information.

(*iii*) **Q-Q Plot:** It graphs the quantiles of one univariate distribution against the corresponding quantiles of another. It is a powerful visualization tool as it allows the user to view whether there is a shift in going from one visualization to another.

(*iv*) **Scatter Plot:** It is one of the most effective graphical methods for determining if there appears to be a relationship, pattern, or trend between two numerical attributes. To construct a scatter plot, each pair of values is treated as a pair of coordinates in an algebraic sense and plotted as points in the plane.

(*v*) **Loess Curve:** A loess curve is another important exploratory graphic aid that adds a smooth curve to a scatter plot in order to provide the better perception of the pattern of dependence.

SUMMARY

- Data quality is an important concept in data mining. The data quality can be measured in accuracy, completeness, consistency, timeliness and interpretability.

- The major tasks involved in data preprocessing are Data Cleaning, Data Integration, Data Reduction and Data Transformation.
- Discretization is a process in which continuous or real valued attributes can be discretized. By discretization we mean that the values of an attribute are partitioned into intervals.
- Concept hierarchy is used to transform the data into multiple level of granularity.
- The requirements of data mining algorithms are efficient sampling methods, storing immediate results while executing queries, capacity to handle large size data and geometric indexes.
- Hence query language has to be designed to support knowledge discovery processes. In the field of association rule mining, query language is required to support the declarative specification of data and pattern manipulations.
- The graphic user interface should be of high quality and should be highly interactive because otherwise it would be much difficult for business users to understand it properly.
- The components of a data warehouse of a mining system are Data Warehouse, Data Warehouse Server, Knowledge Base, Data Mining Engine, Pattern Evolution Module and GUI.
- Data reduction is the step needed when the amount of data selected is very large. The reduction gives a set of data which is very small in volume.

REVIEW QUESTIONS

1. Explain the concept hierarchy process.
2. Why do we require query language?
3. Explain data aggregation.
4. Give an overview of the data mining architecture
5. Write short note on numerosity reduction and data compression.

■■■

CHAPTER 3

Associative Mining Rule

Association rules are one of the most researched areas of data mining and have recently received much attention from the database community. They have proven to be quite useful in the marketing and retail communities as well as other more diverse fields. Association rules, first introduced in 1993, are used to identify relationships among a set of items in a database. These relationships are not based on inherent properties of the data themselves as with functional dependencies, but rather based on co-occurrence of the data items.

Association rules mining is one of the most well studied data mining tasks. It discovers relationships among attributes in databases, producing if-then statements concerning attribute-values. Association rule mining has been applied to web-based education systems from two points of view:

- help faculties to obtain detailed feedback of the *e*-learning process: *e.g.,* finding out how the students learn on the web, to evaluate the students based on their navigation patterns, to classify the students into groups, to restructure the contents of the web site to personalize the courses;
- help students in their interaction with the e-learning system: *e.g.,* adaptation of the course according to the apprentice's progress, *e.g.,* by recommending to them personalized learning paths based on the previous experiences other similar students.

3.1 INTRODUCTION

The Association Rules model is often associated with "market basket analysis", which is used to discover relationships or correlations among a set of items. It is widely used in data analysis for direct marketing, catalog design, and other business decision-making processes. A typical association rule of this kind asserts the likelihood that, for example,"70% of the students who purchase pencil, sharpener, and scale also buy eraser."

Association rules capture the co-occurrence of items or events in large volumes of customer transaction data. Because of progress in bar-code technology, it is now possible for retail organizations to collect and store massive amounts of sales data, referred to as "basket data." Association rules were initially defined on basket data, even though they are applicable in several other applications. Finding all such rules is valuable for cross-marketing and mail-order promotions, but there are other applications as well: catalog design, add-on sales, store layout, customer segmentation, web page personalization, and target marketing.

Traditionally, association rules are used to discover business trends by analyzing customer transactions. However, they can also be used effectively to predict Web page accesses for personalization. For example, assume that after mining the Web access log we discovered an association rule "A and B implies C," with 80% confidence, where A, B, and C are Web page accesses. If a user has visited pages A and B, there is an 80% chance that he/she will visit page C in the same session. Page C may or may not have a direct link from A or B. This information can be used to create a link dynamically to page C from pages A or B so that the user can "click-through" to page C directly. This kind of information is particularly valuable for a Web server supporting an e-commerce site to link the different product pages dynamically, based on the customer interaction.

Association rule mining can be formally defined as follows: Let I = $\{i_1, i_2, ..., i_n\}$ be a set of literals (constants: either a number or a character) called *items* and D be a set of *transactions* where each transaction T is a set of items such that T is a subset of I. Associated with each transaction is an identifier, called its TID. An association rule is an implication of the form X implies Y, where X and Y are both subsets of I, and X intersect Y is empty. The rule has support *s* in the database D if *s*% of the transactions in D contain both X and Y, and confidence *c* if *c*% of transactions that contain X also contain Y. The problem of mining association rules is to generate all rules that have support and confidence greater than the user-specified minimum support and minimum confidence, respectively.

Algorithms that calculate association rules work in two phases. In the first phase, all combinations of items that have the required minimum support (called the "frequent item sets") are discovered. In the second phase, rules of the form X implies Y with the specified minimum confidence are generated from the frequent item sets. Typically the first phase is computationally expensive and has in recent years attracted attention from researchers all over the world. This has resulted in several innovative techniques for discovering frequent item sets.

There are several properties of association rules that can be calculated. ODM supports two:

- **Support:** Support of a rule is a measure of how frequently the items involved in it occur together. Using probability notation, support (A implies B) = P(A, B).
- **Confidence:** Confidence of a rule is the conditional probability of B given A; confidence (A implies B) = P (B given A), which is equal to P(A, B) or P(A).

These statistical measures can be used to rank the rules and hence the predictions.

A formal statement of the association rule problem is given below:

Definition 1: The *support (s)* of an association rule is the ratio (in percent) of the records that contain X″ Y to the total number of records in the database. Therefore, if we say that the support of a rule is 5% then it means that 5% of the total records contain X″ Y.

Support is the statistical significance of an association rule. Grocery store managers probably would not be concerned about how peanut butter and bread are related if less than 5% of store transactions have this combination of purchases. While a high support is often desirable for association rules, this is not always the case. For example, if we were using association rules to predict the failure of telecommunications switching nodes based on what set of events occur prior to failure, even if these events do not occur very frequently association rules showing this relationship would still be important.

Definition 2: For a given number of records, *confidence* (*a*) is the ratio (in percent) of the number of records that contain X″ Y to the number of records that contain X. Thus, if we say that a rule has a confidence of 85%, it means that 85% of the records containing X also contain Y.

The confidence of a rule indicates the degree of correlation in the dataset between X and Y. Confidence is a measure of a rule's strength. Often a large confidence is required for association rules.

If a set of events occur a small percentage of the time before a switch failure or if a product is purchased only very rarely with peanut butter, these relationships may not be of much use for management.

Mining of association rules from a database consists of finding all rules that meet the user-specified threshold support and confidence. The problem of mining association rules can be decomposed into two sub problems as stated in Algorithm 1.

Algorithm 1

Input:

I, D, *s*, *a*

Output:

Association rules satisfying *s* and *a*.

Algorithm:

1. Find all sets of items which occur with a frequency that is greater than or equal to the user-specified threshold support, *s*.
2. Generate the desired rules using the large itemsets, which have user-specified threshold confidence, *a*.

The first step in Algorithm 1 finds *large* or *frequent itemsets*. Itemsets other than those are referred as *small itemsets*. Here an itemset is a subset of the total set of items of interest from the database. An interesting (and useful) observation about large itemsets is that:

If an itemset X is small, any superset of X is also small. Of course the contra-positive of this statement (If X is a large itemset than so is any subset of X) is also important to remember. We use L to designate the set of large itemsets. The second step in Algorithm 1 finds association rules using large itemsets obtained in the first step.

Algorithm 2. Find Association Rules Given Large Itemsets:

Input:

I, D, *s*, *a*, L

Output:

Association rules satisfying *s* and *a*.

Algorithm:

1. Find all nonempty subsets, *x*, of each large itemset, $l \in L$
2. For every subset, obtain a rule of the form $x => (l-x)$ if the ratio of the frequency of occurrence of l to that of x is greater than or equal to the threshold confidence.

2. Apriori Algorithm-Single-Dimensional Boolean Association Rules from Transactional Databases

Apriori is a classic algorithm for learning association rules. Apriori is designed to operate on databases containing transactions, for example, collections of items bought by customers, or details of a website frequentation. Other algorithms are designed for finding association rules in data having no transactions (Winepi and Minepi), or having no timestamps (DNA sequencing).

As is common in association rule mining, given a set of *itemsets* (for instance, sets of retail transactions, each listing individual items purchased), the algorithm attempts to find subsets which are common to at least a minimum number C of the itemsets. Apriori uses a "bottom up" approach, where frequent subsets are extended one item at a time, a step known as *candidate generation*, and groups of candidates are tested against the data. The algorithm terminates when no further successful extensions are found.

Apriori uses breadth-first search and a tree structure to count candidate item sets efficiently. It generates candidate item sets of length k from item sets of length $k - 1$. Then it prunes the candidates which have an infrequent sub pattern. According to the downward closure lemma, the candidate set contains all frequent k-length item sets. After that, it scans the transaction database to determine frequent item sets among the candidates.

Apriori, while historically significant, suffers from a number of inefficiencies or trade-offs, which have spawned other algorithms. Candidate generation generates large numbers of subsets (the algorithm attempts to load up the candidate set with as many as possible before each scan). Bottom-up subset exploration (essentially a breadth-first traversal of the subset lattice) finds any maximal subset S only after all $2^{|S|} - 1$ of its proper subsets.

The major steps in association rule mining are:

1. Frequent Itemset generation
2. Rules derivation

The APRIORI algorithm uses the downward closure property, to prune unnecessary branches for further consideration. It needs two parameters, minSupp and minConf. The minSupp is used for generating frequent itemsets and minConf is used for rule derivation.

The APRIORI algorithm:

1. $k = 1$;
2. Find frequent itemset, L_k from C_k, the set of all candidate itemsets;
3. Form C_{k+1} from L_k;
4. $k = k + 1$;
5. Repeat 2–4 until C_k is empty;

Step 2 is called the frequent itemset generation step. Step 3 is called as the candidate itemset generation step. Details of these two steps are in the next lesson.

Frequent itemset generation: Scan D and count each itemset in C_k, if the count is greater than minSupp, then add that itemset to L_k.

Candidate itemset generation

For $k = 1$, C_1 = all itemsets of length = 1.

For $k > 1$, generate C_k from L_{k-1} as follows:

The join step:

$C_k = k - 2$ way join of L_{k-1} with itself.

If both $\{a_1, \ldots, a_{k-2}, a_{k-1}\}$ and $\{a_1, \ldots, a_{k-2}, a_k\}$ are in L_{k-1}, then add $\{a_1, \ldots, a_{k-2}, a_{k-1}, a_k\}$ to C_k.

The items are always stored in the sorted order.

The prune step:

Remove $\{a_1, \ldots, a_{k-2}, a_{k-1}, a_k\}$, if it contains a non-frequent $(k-1)$ subset.

3.2 MULTI-LEVEL AND MULTI-DIMENSIONAL ASSOCIATION RULES FROM TRANSACTION DATABASES

So far we have studied association rules that imply a single predicate, that is, the predicate buys. We refer to each distinct predicate in a rule as a dimensional. Hence, we can refer to the above rule as a single-dimensional or intra-dimension association rule since it contains a single distinct predicate (*e.g.*, buys) with multiple occurrences. (*i.e.*, predicate occurs more than once within the rule). Such rules are commonly mined from transactions data. Suppose that rather than using a transactional database, sales and related information are stared in a relational database or data warehouse. Such data stares are multidimensional, by definition. For instance, in addition to keeping track of the items purchased in sales transactions, a relational database may record other attributes associated with the items, such as the quantity purchased or the price, or the branch location of the sale. Additional relational information regarding the customers who purchased the items, such as customer are, occupation, credit rating, income, and address, may also be stored. Considering each database attribute or warehouse dimension as a predicate, it can therefore be interesting to mine association rules containing multiple predicates, such as

age(*x*, "20......29") ^ occupation (X, "student") implies buys (X, "laptop":)

Association rules that involve two or more dimensions or predicates can be referred to as multidimensional association rules. The above rule contains three predicates (age, occupation, and buys), each of which occurs only once in the rule. Hence, we say that it has no repeated predicates. Multidimensional association rules with no repeated predicates are called inter-dimensional association rules. We may also be interested in mining multidimensional association rules with repeated predicates, which contain multiple occurrences of some predicates. These rules are called hybrid-dimension association rules. An example of such a rule is the following, where the predicate buys is repeated:

Age (X, "20......29") ^ buys (X, "laptop") implies buys (X, "sony b/w printer")

Note that database attributes can be categorical or quantitative. Categorical attributes have a finite number of possible values, with no ordering among the values (*e.g.*, occupation, brandy color). Categorical attributes are also called nominal attributes, since their values are "names of things "Quantitative attributes are numeric and have an implicit ordering among values (*e.g.* age, income, price). Techniques for mining multidimensional association rules can be categorized according to three basic approaches regarding the treatment of quantitative attributes.

In the first approach, quantitative attributes are discretized using predefined concept hierarchies. This discretization occurs prior to mining. Here, discretization is static and predetermined. The discretized numeric attributes, with their range values, can then be treated as categorical attributes (where each range is considered a category). We refer to this as mining multidimensional association rules using static discretization of quantitative attributes.

In the second approach, quantitative attributes are discretized into "bins" based on the distribution of the data. These bins may be further combined during the mining process. The discretization process is dynamic and established so as to satisfy some mining criteria, such as maximizing the confidence of the rules mined. Because this strategy treats the numeric attribute values as quantities rather than as predefined ranges or categories, association rules mined from this approach are also referred to as quantitative association rules.

In the third approach, quantitative attributes are discretized so as to capture the semantic meaning of such interval data. This dynamic discretization procedure considers the distance between data points. Hence, such quantitative association rules are also referred to as distance-based association rules.

For many applications, it is difficult to find strong associations among data items at low or primitive levels of abstraction due to the sparsity of data in multidimensional space. Strong associations discovered at high concept levels that might represent common sense knowledge. However, what may represent common sense to one user may seem novel to another. Therefore, data mining systems should provide capabilities to mine association rules at multiple levels of abstraction and traverse easily among different abstraction spaces.

Approaches to mining multilevel Association Rules: There are some approaches based on a support-confidence framework. In general, a top-down strategy is employed, where counts are accumulated for the calculation of frequent itemsets at each concept level, starting at the concept level 1 and working towards the lower, more specific concept levels, until no more frequent itemsets can be found. That is, once all frequent itemsets at concept level 1 are found, then the frequent itemsets at level 2 are found, and so on. For each level, any algorithm for discovering frequent itemsets may be used, such as Apriori or its variations.

- Using uniform minimum support for all levels (referred to as uniform support). The same minimum support threshold is used when mining at each level of abstraction.

When a uniform minimum support threshold is used, the search procedure is simplified. The method is also simple in that users are required to specify only one minimum support threshold. An optimization technique can be adopted, based on the knowledge that an ancestor is a superset of its descendents; the search avoids examining itemsets containing any item whose ancestors do not have minimum support. The uniform support approach, however, has some difficulties. It is unlikely that items at lower levels of abstraction will occur as frequently as those at higher levels of abstraction. If the minimum support threshold is set too high, it could miss several meaningful associations occurring at low abstraction levels. If the threshold is set too low, it may generate many uninteresting associations occurring at high abstraction levels. This provides the motivation for the following approach.

Using reduced minimum support at lower levels (referred to as reduced support): Each level of abstraction has its own minimum support threshold. The lower the abstraction level, the smaller the corresponding threshold.

For mining multiple-level associations with reduced support, there are a number of alternative search strategies:

Level-by-Level independent: This is a full-breadth search, where no background knowledge of frequent itemsets is used for pruning. Each node is examined, regardless of whether or not its parent node is found to be frequent.

Level-cross-filtering by single item: An item at the ith level is examined if and only if its parent node at the $(i-1)$th level is frequent. In other words, we investigate a more specific association from a more general one. If a node is frequent, its children will be examined; otherwise, its descendents are pruned from the search.

Level-cross filtering by -K-itemset: A A-itemset at the ith level is examined if and only if its corresponding parent A-itemset at the $(i-1)$th level is frequent.

Advantages:

➢ Search procedure is simplified.

➢ Avoids examining itemsets containing any item whose ancestors do not have minimum support.

Disadvantages:

- It is unlikely that items at lower levels of abstraction will occur as frequently as those at higher levels of abstraction.
- If the minimum support threshold is set too high, it could miss some meaningful associations occurring at low abstraction levels.
- If the threshold is set too low, it may generate many uninteresting associations occurring at high abstraction levels.

Using item or group-based minimum support (referred to as group-based support): Because users or experts often have insight as to which groups are more important than others, it is sometimes more desirable to set up user-specific, item, or group based minimal support thresholds when mining multilevel rules. For example, a user could set up the minimum support thresholds based on product price, or on items of interest, such as by setting particularly low support thresholds for laptop computers and flash drives in order to pay particular attention to the association patterns containing items in these categories.

3.3 ASSOCIATION MINING TO CORRELATION ANALYSIS

Correlation analysis is very important in selecting variables for clustering/segmentation and predictive modeling. Correlation coefficient is a numerical measurement of linear association between two numerical variables. Correlation coefficients range between −1 and 1. If two variables are perfectly negatively correlated, the coefficient is −1. If two are perfectly positively correlated, it is 1. Simple coefficient computation reveals linear correlations as shown below:

Figure 3.1. Linear Correlation

However, the following type correlation cannot be exposed with simple computation. To compute non-linear correlation, CMSR employs advanced techniques to identify non-linear correlation.

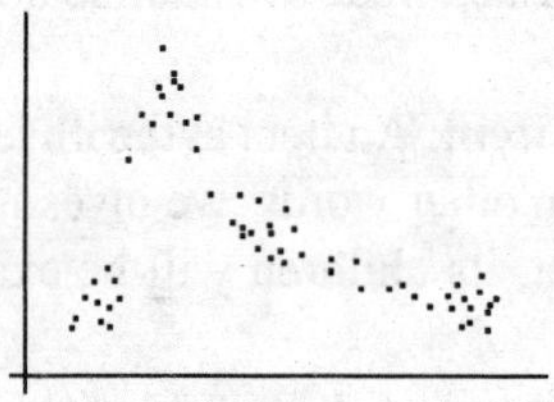

Figure 3.2. Non-Linear Correlation

In addition, correlation coefficients cannot be computed directly from categorical variables. Normally, linearization techniques are used.

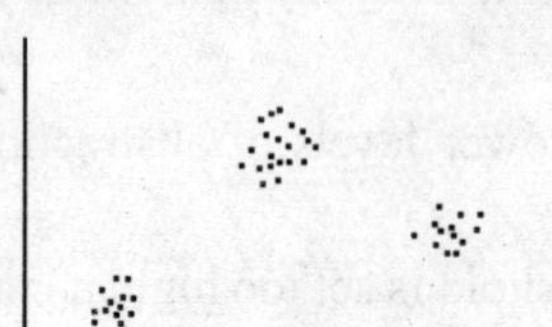

Figure 3.3. Non-linear Categorical Correlation

3.4 CONSTRAINT BASED ASSOCIATION MINING

A data mining process may uncover thousands of rules from a given set of data, most of which end up being unrelated or uninteresting to the users. A good heuristic is to have the users specify such intuition or expectations as constraints to confine the search space. This strategy is known as constraint-based mining. The constraints can include the following:

- **Knowledge type constraints:** These specify the type of knowledge to be mined, such as association or correlation.
- **Data constraints:** These specify the set of task-relevant data.
- **Dimension/level constraints:** These specify the desired dimensions (or attributes) of the data, or levels of the concept hierarchies, to be used in mining.
- **Interestingness constraints:** These specify thresholds on statistical measures of rule interestingness, such as support, confidence, and correlation.
- **Rule constraints:** These specify the form of rules to be mined. Such constraints may be expressed as metarules (rule templates), as the maximum or minimum number of predicates that can occur in the rule antecedent or consequent, or as relationships among attributes, attribute values, and/or aggregates.

The above constraints can be specified using a high-level declarative data mining query language and user interface.

1. Metarule-Guided Mining of Association Rules

- Metarules allow users to specify the syntactic form of rules that they are interested in mining. The rule forms can be used as constraints to help improve the efficiency of the mining process.
- Metarules may be based on the analyst's experience, expectations, or intuition regarding the data or may be automatically generated based on the database schema.
- **Example:** Suppose one is interested in finding associations between customer traits and the items that customers buy-determining which pairs of customer traits promote the sale of office software.
- Meta rule will be:

$$P1(X, Y) \wedge P2(X,W) \rightarrow buys(X, \text{“office software”})$$

where P1 and P2 are predicate variables that are instantiated to attributes from the given database during the mining process, X is a variable representing a customer, and Y and W take on values of the attributes assigned to P1 and P2, respectively. Typically, a user will specify a list of attributes to be considered for instantiation with P1 and P2.

- The data mining system can then search for rules that match the given metarule. For instance, this rule matches or complies with earlier metarule.

$$age(X, \text{“30...39”}) \wedge income(X, \text{“41K...60K”}) \rightarrow buys(X, \text{“office software”})$$

Metarules to guide mining process:

- A metarule is a rule template of the form

$$P_1 \wedge P_2 \wedge ... \wedge P_1 \rightarrow Q1 \wedge Q2 \wedge .... \wedge Qr$$

- where P*i* (*i* = 1, ..., l) and Q*j* (*j* = 1, ..., *r*) are either instantiated predicates or predicate variables.
- Let the number of predicates in the metarule be $p = l + r$.
- To find interdimensional association rules satisfying the template,

➢ find all frequent *p*-predicate sets, L*p*.

➢ have the support or count of the l-predicate subsets of L*p* in order to compute the confidence of rules derived from L*p*.

2. Constraint Pushing: Mining Guided by Rule Constraints

- Rule constraints specify expected set/subset relationships of the variables in the mined rules, constant initiation of variables, and aggregate functions.
- These rule constraints may be used with metarule-guided mining.
- For example, multidimensional database with the following interrelated relations:
  ➢ Dimension Tables:
    - sales (customer name, item name, TID)
    - lives in (customer name, region, city)
    - item (item name, group, price)
    - transaction (TID, day, month, year)

  ➢ Fact table sales:
    - customer name, item name, and TID (transaction id).
- Mining query is: "Find the sales of which cheap items (where the sum of the prices is less than $100) may promote the sales of which expensive items (where the minimum price is $500) of the same group for Chicago customers in 2004."
- This can be expressed in DMQL:
  1. mine associations as
  2. lives in(C, , "Chicago") ^ sales $^+$(C,?{I},{S}) sales$^+$(C, ,?{I},{S})
  3. from sales
  4. where S.year = 2004 and T.year = 2004 and I.group = J.group
  5. group by C, I.group
  6. having sum(I.price) < 100 and min(J.price) > 500
  7. with support threshold = 1%
  8. with confidence threshold = 50%
- Line 1 is a knowledge type constraint, where association patterns are to be discovered. Line 2 specifies a metarule. Dimension/level constraints and interestingness constraints can be applied after mining to filter out discovered rules, although it is generally more efficient and less expensive to use them during mining, to help prune the search space.

Classification of Rule constraints: Rule constraints can be classified into the following five categories with respect to frequent itemset mining: (1) antimonotonic, (2) monotonic, (3) succinct, (4) convertible, and (5) inconvertible.

(*a*) **Antimonotonic:**

- If an itemset does not satisfy this rule constraint, none of its supersets can satisfy the constraint. If a rule constraint obeys this property, it is antimonotonic.
- "avg (I. price) <= 100" is not antimonotonic.
- For a given itemset that does not satisfy this constraint, a superset created by adding some (cheap) items may result in satisfying the constraint.

- Pruning by antimonotonic constraints can be applied at each iteration of Apriori-style algorithms to help improve the efficiency of the overall mining process while guaranteeing completeness of the data mining task.

(*b*) **Monotonic:**

- Of an itemset satisfies this rule constraint, so do all of its supersets. If a rule constraint obeys this property, it is monotonic.
- Example: "sum(I. price) > 100,"
- If an itemset I satisfies the constraint, that is, the sum of the prices in the set is no less than 100, further addition of more items to I will increase cost and will always satisfy the constraint.

(*c*) **Succinct:**

- All and only those set guaranteed to satisfy the rule can be enumerated.
- The itemsets can be generated that satify the rule even before support counting begins.
- Once such subset is generated, iterative testing for the constraint can be effectively avoided. For example, the constraint "min(J. price) > 500", "Max(S) < 120"
- avg(S) > *v*, avg(S) < *v* are not succinct.

(*d*) **Convertible:**

- If the items in the itemset are arranged in a particular order, the constraint may become monotonic or antimonotonic with regard to the frequent itemset mining process.
- For example, the constraint "avg(I. price) <= 100" is neither antimonotonic nor monotonic.
- However, if items in a transaction are added to an itemset in price-ascending order, the constraint becomes antimonotonic, because if an itemset I violates the constraint (*i.e.*,with an average price greater than $100), then further addition of more expensive items into the itemset will never make it satisfy the constraint.

(*e*) **Inconvertible:**

- Constraints which are not convertible.
- For example, "sum(S) < *v*,", "sum(S) < V," where element in S could be of any real value, is not convertible.

3.5 TASK AND NAIVE ALGORITHM

Various tasks in data mining are:

Classification is one of the most popular **data mining tasks**. Business problems like churn analysis, risk management and ad targeting usually involve classification. There are some business problems for data mining base on its company. Classification refers to assigning cases into categories based on a predictable attribute. Each case contains a set of attributes, one of which is the class attribute (predictable attribute). The task requires finding a model that describes the class attribute as a function of input attributes. **Data mining** algorithms that require a target to learn against are considered supervised algorithms. Typical classification algorithms include decision trees, neural network in database, and Naïve Bayes.

Clustering is also called segmentation. It is used to identify natural groupings of cases based on a set of attributes. Cases within the same group have more or less similar attribute values. To avoid large amount of data, sometimes data mining vendor and product do a database archiving. Figure below displays a simple customer dataset containing two attributes: age and income. The clustering algorithm groups the dataset into three segments based on these two attributes.

Cluster 1 contains the younger population with a low income. Cluster 2 contains middle-aged customers with higher incomes. Cluster 3 is a group of senior individuals with a relatively low income. Clustering is an unsupervised **data mining tasks**. No single attribute is used to guide the training process. All input attributes are treated equally. Most clustering algorithms build the model through a number of iterations and stop when the model converges, that is, when the boundaries of these segments are stabilized.

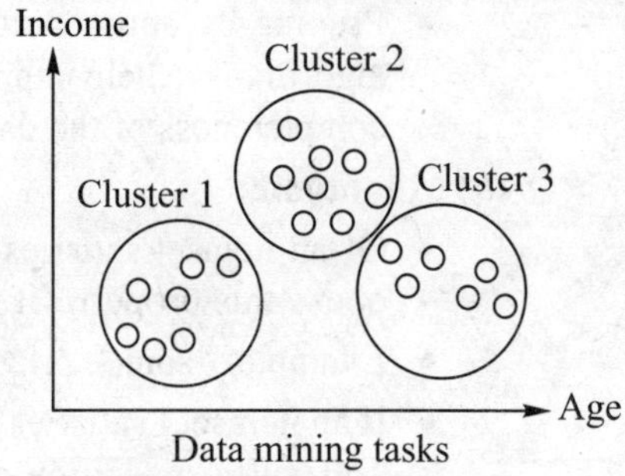

Figure 3.4. Data Mining Tasks

Association

Association is another popular **data mining task**. Association is also called market basket analysis. A typical association business problem is to analyze a sales transaction table and identify those products often sold in the same shopping basket. The common usage of association is to identify common sets of items (frequent itemsets) and rules for the purpose of cross-selling.

Association is interaction and data communication to provide useful information. In terms of association, each product, or more generally, each attribute/value pair is considered an item. The association task has two goals: to find frequent itemsets and to find association rules.

Most **association** type algorithms find frequent itemsets by scanning the dataset multiple times. The frequency threshold (support) is defined by the user before processing the model. For example, support = 2% means that the model analyzes only items that appear in at least 2% of shopping carts. A frequent itemset may look like {Product = "Pepsi", Product = "Chips", Product = "Juice"}. Each itemset has a size, which is the number of items that it contains. The size of this particular itemset is 3.

Association algorithm is very dependent on database design and construction, how important database design is depend on its size of data. Some of them use distributed database to avoid the crash of DBMS. Apart from identifying frequent itemsets based on support, most association type algorithms also find rules. An association rule has the form A, B => C with a probability, where A, B, C are all frequent item sets. The probability is also Cluster 2, Cluster 3, Cluster 1, Age, Income referred to as the confidence in data mining literature.

The probability is a threshold value that the user needs to specify before training an association model. For example, the following is a typical rule: Product = "Pepsi", Product = "Chips" => Product = "Juice" with an 80% probability. The interpretation of this rule is straightforward. If a customer buys Pepsi and chips, there is an 80% chance that he or she may also buy juice. Figure 3.4 displays the product association patterns. Each node in the figure represents a product, each edge represents the relationship. This concept is similar with Entity Relationship Diagram. The direction of the edge represents the direction of the prediction. For example, the edge from Milk to Cheese indicates that those who purchase milk might also purchase cheese.

Regression

The **regression task** is similar to classification. The main difference is that the predictable attribute is a continuous number. Regression techniques have been widely studied for centuries in the field of statistics. Linear regression and logistic regression are the most popular regression methods. Other regression techniques include regression trees and neural networks.

Regression tasks can solve many business problems. For example, they can be used to predict coupon redemption rates based on the face value, distribution method, and distribution volume, or to predict wind velocities based on temperature, air pressure, and humidity. Data mining is a large of data, so when it collected in database, it must use indexing technique for tuning database.

Forecasting

Forecasting is yet another important **data mining task**. Database Development in globalization era is the cause a useful data mining algorithm. What will the sales amount of Pepsi be next month? Forecasting can help to answer these questions. It usually takes as an input time series dataset, for example a sequence of numbers with an attribute representing time. The time series data typically contains adjacent observations, which are order-dependant. Forecasting techniques deal with general trends, periodicity, and noisy noise filtering. The most popular time series technique is ARIMA, which stands for Auto Regressive Integrated Moving Average model.

Figure 3.5 contains two curves. The solid line curve is the actual time series data on Microsoft stock value, while the dotted curve is a time series model based on the moving average forecasting technique.

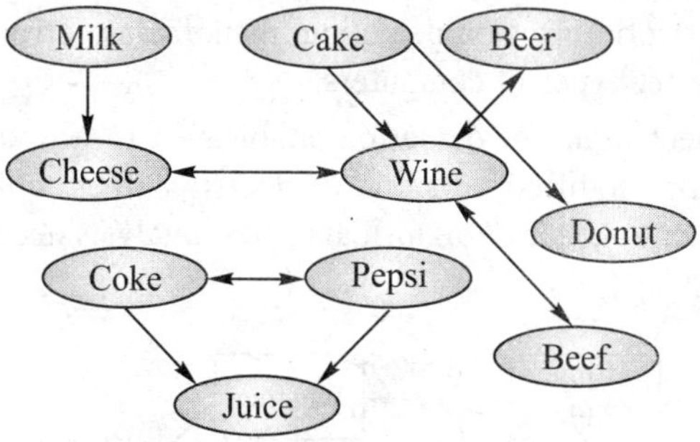

Product Association

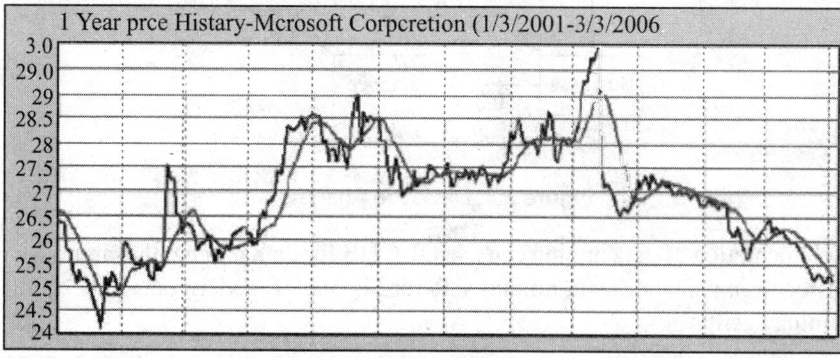

Times Series

Figure 3.5. Forecasting

Sequence Analysis

Computer development is have contribute to data mining algorithm. Sequence analysis is used to find patterns in a discrete series. A sequence is composed of a series of discrete values (or states). For example, a DNA sequence is a long series composed of four different states: A, G, C, and T. A Web click sequence contains a series of URLs. Customer purchases can also be modeled as sequence data. For example, a customer first buys a computer, then speakers, and finally a Webcam. Both sequence and time series data contain adjacent observations that are dependant. The difference is that the sequence series contains discrete states, while the time series contains continuous numbers. **Data mining tasks** of sequence analysis is very complex.

Sequence and association data are similar in the sense that each individual case contains a set of items or states. The difference between sequence and association models is that sequence models analyze the state transitions, while the association model considers each item in a shopping cart to be equal and independent. With the sequence model, buying a computer before buying Cheese, Wine, Milk, Cake, Beer, Coke, Pepsi, Juice, Beef, and Donut speakers is a different sequence than buying speakers before a computer. With an **association algorithm**, these are considered to be the same itemset.

Deviation Analysis

Deviation analysis is for finding those rare cases that behave very differently from others. It is also called outlier detection, which refers to the detection of significant changes from previously observed behaviour. Deviation analysis can be used in many applications. The most common one is credit card fraud detection. To identify abnormal cases from millions of transactions is a very challenging task. Other applications include network intrusion detection, manufacture error analysis, and so on. Using this algorithm is dependent on the types of computers.

There is no standard technique for deviation analysis. It is still an actively researched topic. Usually analysts employ some modified versions of decision trees, clustering, or neural network algorithms for this task. In order to generate significant rules, analysts need to oversample the anomaly cases in the training dataset.

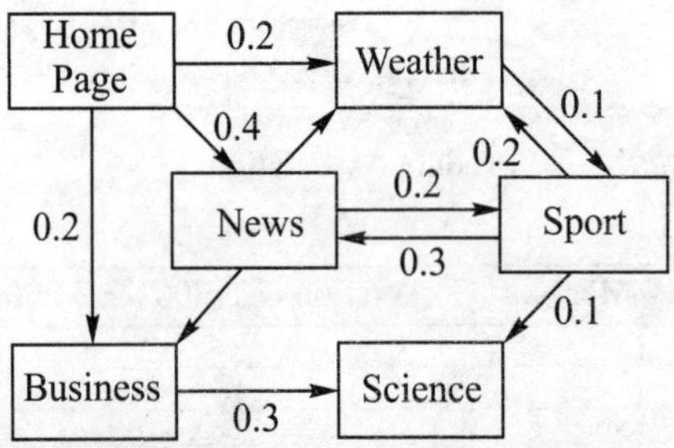

deviation analysis data mining tasks

Figure 3.6. Deviation Analysis

The implementation of data mining can use OLE DB for data mining, this approach could make data mining easier to implement. Data mining software is one of modern computer applications that support data mining easily.

Naïve algorithms include apriori algorithm series that we will discuss in the coming sections.

3.6 IMPROVING THE EFFICIENCY OF APRIORI ALGORITHM

The variations of Apriori to improve the efficiency of the original algorithm are:

1. **Hash-based technique (hashing itemsets into corresponding buckets):** A *k*-itemset whose corresponding hashing bucket count is below the threshold cannot be frequent.

2. **Transaction reduction (reducing the number of transactions scanned in future iterations):** A transaction that does not contain any frequent k-itemsets is useless in subsequent scans.
3. **Partitioning (partitioning the data to find candidate itemsets):** Any itemset that is potentially frequent with respect to database must occur as a frequent itemset in at least one of the partitions.
4. **Sampling (mining on a subset of the given data):** Mining of a subset of given data, lower support threshold + a method to determine the completeness.
5. **Dynamic itemset counting (adding candidate itemsets at different points during a scan):** Add a new candidate itemsets only when all of their subsets are estimated to be frequent.

One way to improve efficiency of the APRIORI would be to prune without checking all k–1 subsets and join without looping over the entire set, L_{k-1}

This can be done by using *hash trees*. Other methods to improve efficiency are:

- Speed up searching and matching.
- Reduce the number of transactions (a kind of instance selection).
- Reduce the number of passes over data on disk. *E.g. Reducing scans via Partition*.
- Reduce number of subsets per transaction that must be considered.
- Reduce number of candidates (a kind of feature selection).

Hash Trees

A *Hash tree* stores all candidate k-itemsets and their counts. The root is empty and its children are the frequent 1-itemsets. Any node at depth = k will denote and frequent k-itemset. An example for an hash tree for C_2 = 12, 13, 15, 23, 25, 35 is shown below:

```
                        { }                          ** root
          /1             \2          \3              ** edge+label
    /2    |3    \ 5     / 3   \ 5    / 5
  [12:] [13:]  [15:]  [23:] [25:]   [35:]            ** leaves
```

An internal node v at level m contains, bucket pointers. These tell which branch is the next one to be traversed. The hash of the mth item is used to decide this.

Join step using Hash Tree

Only the frequent k–1 itemsets, who have common parents should be considered for the joining step. So checking all k–1 itemsets in $L_{k-1}$is avoided.

Prune step using Hash Tree

To determine if a k–1 itemset is frequent, we have to look only for those itemsets who have common parents, and thus avoid going through all k–1 itemsets in L_{k-1}.

Added advantages of Hash Trees

Enumeration is replaced by tree traversal, *i.e.*, there is no need to enumerate all *k*-subsets of transactions. Such considerations are limit by tree traversal.

3.7 APRIORI-TID

With large databases, it is required to have algorithms that process the data with high speed. For discovering association rules between items in a large database, we have had algorithms like AIS algorithm and SETM algorithm but the speed with which they processed the data in the databases was not fast and also they tend to find a lot of itemsets which were small and thus end up wasting time. The Apriori algorithm makes multiple passes over the database. In the first pass individual item support is counted and the items that have support greater than or equal to minimum support are considered as large items. For each pass k after that, the large itemsets in the previous pass is grouped in sets of k items and these form candidate itemsets. The support for the different candidate itemsets is counted and if the support is found to be greater than minimum support then that itemset is considered to be large. This process continues until the large itemset in a particular pass comes out to be an empty set. Similar to the Apriori algorithm, the AprioriTid algorithm also uses the apriori-gen function to determine the candidate itemsets but the difference is that the database is not used for counting support after the first pass. Instead set of candidate itemsets is used for this purpose for $k > 1$. In case a transaction does not have any candidate k-itemset then the set of candidate itemsets would not have any entry for that transaction which will eventually decrease the number of transaction in the set containing the candidate itemsets as compared to the database. As value of k increases each entry will be smaller than the corresponding transactions because the number of candidates in the transactions will decrease. Apriori performs better than AprioriTid in the initial passes but in the later passes AprioriTid has better performance than Apriori. Due to this reason we can use another algorithm called Apriori Hybrid algorithm in which Apriori is used in the initial passes but we switch to AprioriTid in the later passes.

We will use the apriori-gen function to determine the candidate itemsets before the pass begins. The interesting feature is that the database D is not used for counting support after the first pass. Rather, the set | C*k* | is used for this purpose. Each member of the set | C*k* | is of the form < TID; {X*k*}>, where each X*k* is a potentially large *k*-itemset present in the transaction with identifier TID. For $k = 1$, C1 corresponds to the database D, although conceptually each item *i* is replaced by the itemset {*i*}. For $k > 1$, | C*k* | is generated by the algorithm. The member of C*k* corresponding to transaction *t* is < *t*.TID, {*c*Î C*k* | *c* contained in *t*}>. If a transaction does not contain any candidate *k*-itemset, then | C*k* | will not have an entry for this transaction. Thus, the number of entries in | C*k* | may be smaller than the number of transactions in the database, especially for large values of *k*. In addition, for large values of *k*, each entry may be smaller than the corresponding transaction because very few candidates may be contained in the transaction. However, for small values for *k*, each entry may be larger than the corresponding transaction because an entry in C*k* includes all candidate *k*-itemsets contained in the transaction.

Algorithm AprioriTid

```
(1) L1 = {large 1-itemsets};
(2) C1 = database D;
(3) for ( k = 2; Lk-1 '"Æ ; k++ ) do begin
(4) Ck = apriori-gen(Lk-1);
(5) |Ck| = Æ;
(6) forall entries t Î |Ck-1| do begin
(7) // determine candidate itemsets in Ck contained in the transaction
with identifier t.TID
Ct = {c 2ÎCk | (c - c[k]) Î t.set-of-itemsets ^ (c - c[k-1]) Î t.set-of-
itemsets};
(8) forall candidates c Î Ct do
```

Part 2: From the hash table we'll obtain the set of candidates. These candidates will be examined to generate L_k. When we've finished making L_k, database will be trimmed to remove unuseful items and hash table for next pass will be built.

Part 3: Do the same thing as in part 2, except building hash table.

From the beginning of this part, the significantly difference of candidates and really large itemsets in some first pass, after that, the difference is not too much. Whereas, to create a hash table we must do some extra work, it's not a smart idea if the fraction is higher than one threshold. Therefore, the process contained two different parts, one is used at first, and the other is used when the difference of the candidates and the large itemsets is not much (this threshold depends on the manager).

Pruning task consists of transactions pruning and items pruning: one transaction contained a large $(k+1)$ k-itemsets if it has at least $(k+1)$ large k-itemsets. It's mean that we are able to cut off the transactions which don't have enough $(k+1)$ large k-itemsets. In addition, we found that, if an item belongs to a $(k+1)$ frequent $(k+1)$-itemsets, then, it's contained in at least k frequent k-itemsets $(k+1$ minus 1). Thus, we'll count and trim the items which have the number of appearance in sets of L_k less than k.

Pseudo-code

Main program:

```
/* Part 1*/
s = a minimum support;
Set all the buckets of H2 to zero; //hash table for all transaction t " D
do begin
insert and count 1-items occurrences in a hash tree;
for all 2-subsets x of t do
    H2[h2(x)] ++;
end
L1 = {c|c.count >= s, c is in a leaf node of the hash tree};
/* Part 2 */
k = 2
Dk = D; //DB for large k-itemsets
while (|{x|Hk[x] >= s}| >= LARGE) {
    //make a hash table
    gen_candidate(Lk-1, Hk, Ck);
    set all the buckets of Hk+1 to rezo;
    Dk+1 = ";
for all transactions t " Dk do begin
    count_support(t,Ck,k,t'); // t'" t
    if (|t'| >k) then Dk+1 = Dk+1 *" {t'};
end
Lk = {c " Ck | c.count >= s};
    k++;
    {
/* Part 3 */
gen_candidate (Lk-1, Hk, Ck);
```

```
while (|Ck| > 0) {
    Dk+1 = ";
for all transactions t " Dk
do begin
    count_support(t, Ck,k,t'); // t'" t
If (|t'| > k) then Dk+1 = Dk+1 *" {t'};
end
Lk = {c " Ck | c.count >= s};
if (|Dk+1 | = 0) then break;
Ck+1 = Apriori_gen(Lk);
k++;
    {
Answer = *"k Lk;
Sub procedures:
Procedure gen_candidate (Lk-1, Hk, Ck)
    Ck = ";
    for all c = cp[1] …. Cp[k-2].cp[k-1].cq[k-1], cp,cq " Lk-1, cp )"
                                                    cq = " do
    if (Hk[hk(c)] >= s) then
    Ck = Ck *" {c};
end Procedure
Procedure count_support(t, Ck, k, t)
for all c such that c" Ck and c (= ti1…tik) " t
do
begin
    c.count++;
    for (j = 1; j<= k; j++) a[i
j
    ;++[
end
for (i = 0; j = 0; i < |t|; i++)
    if (a[i] >= k) then do begin t'j
    = t
i
    ; j++; end
end Peocedure
Procedure make_hasht(t', Hk, k, Hk+1, t")
for all (k+1)-subsets x (= t'i1, … , t'ik) of t' do
    if (for all k-subsets y of x, Hk[hk(y)] >= s) then do
    begin
    Hk+1[hk+1(x)]++;
    for (j = 1; j<= k+1; j++) a[i
```

```
j
    ;++[ 
    end
for (j = 11 j <= k+1; j++)
if (a[i] > 0) then do begin t"j = t'i
    ; j++; end
end Procedure
```

3.9 DYNAMIC ITEMSET COUNTING (DIC)

A dynamic itemset counting technique was proposed in which the database is partitioned into blocks marked by start points. In this variation, new candidate itemsets can be added at any start point, unlike in Apriori, which determines new candidate itemsets only immediately prior to each complete database scan. The techniqueis dynamic in that it estimates the support of all of the itemsets that have been counted so far, adding newcandidate itemsets if all of their subsets are estimated to be frequent. The resulting algorithm requires twodatabase scans.

DIC Algorithm

Algorithm:

1. Mark the empty itemset with a solid square. Mark all the 1-itemsets with dashed circles. Leave all other itemsets unmarked.
2. While any dashed itemsets remain:
    1. Read *M* transactions (if we reach the end of the transaction file, continue from the beginning). For each transaction, increment the respective counters for the itemsets that appear in the transaction and are marked with dashes.
    2. If a dashed circle's count exceeds *minsupp*, turn it into a dashed square. If any immediate superset of it has all of its subsets as solid or dashed squares, add a new counter for it and make it a dashed circle.
    3. Once a dashed itemset has been counted through all the transactions, make it solid and stop counting it.

Itemset lattices: An itemset lattice contains all of the possible itemsets for a transaction database. Each itemset in the lattice points to all of its supersets. When represented graphically, a itemset lattice can help us to understand the concepts behind the DIC algorithm.

- Example: *minsupp* = 25% and M = 2.

TID	A	B	C
T1	1	1	0
T2	1	0	0
T3	0	1	1
T4	0	0	0

Transaction Database

Itemset lattice for the above transaction database:

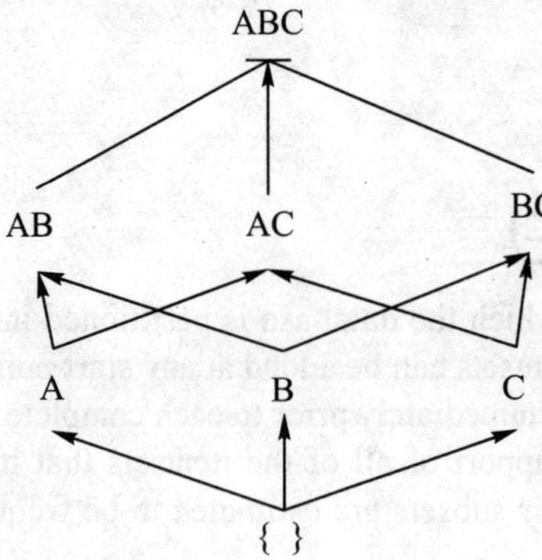

Itemset lattice before any transactions are read:

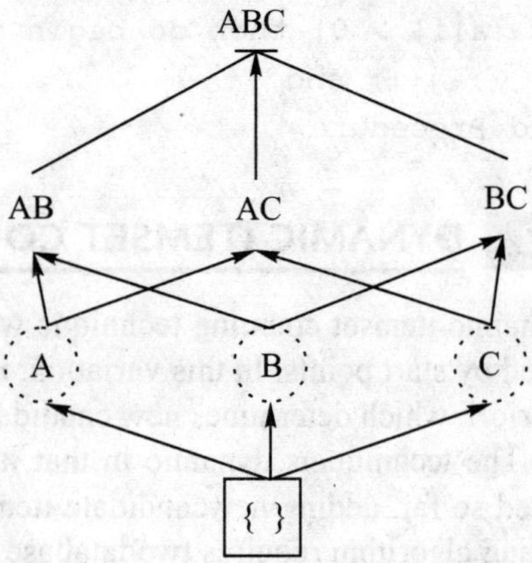

Counters: A = 0, B = 0, C = 0 Empty itemset is marked with a solid box. All 1-itemsets are marked with dashed circles.

After M transactions are read:

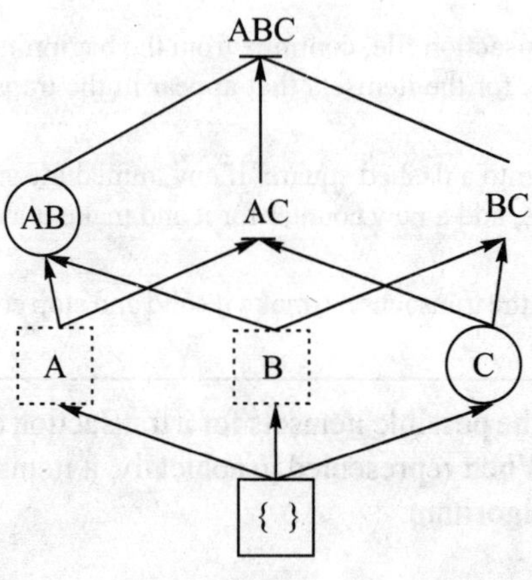

Counters: A = 2, B = 1, C = 0, AB = 0. We change A and B to dashed boxes because their counters are greater than minsup (1) and add a counter for AB because both of its subse*t*s are boxes.

After 2M transactions are read:

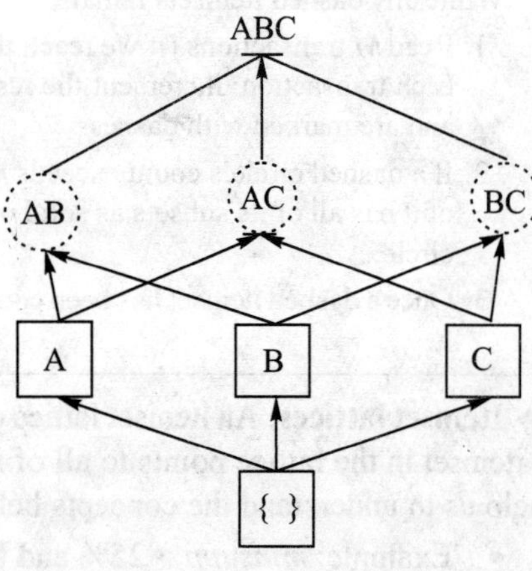

Counters: A = 2, B = 1, C = 0, AB = 0. C changes to a square because its counter is greater than minsup.A, B and C have been counted all the way through so we stop counting them and make their boxes solid. Add counters for AC and BC because their subsets are all boxes.

After 3M transactions read:

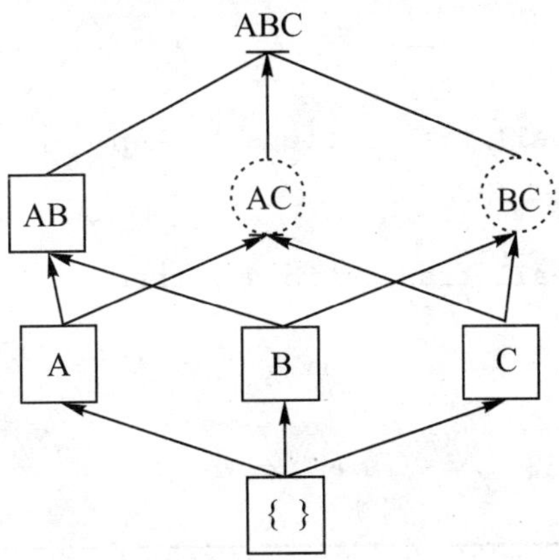

Counters: A = 2, B = 2, C = 1, AB = 1, AC = 0, BC = 0. AB has been counted all the way through and its counter satisfies minsup so we change it to a solid box. BC changes to a dashed box.

After 4M transactions read:

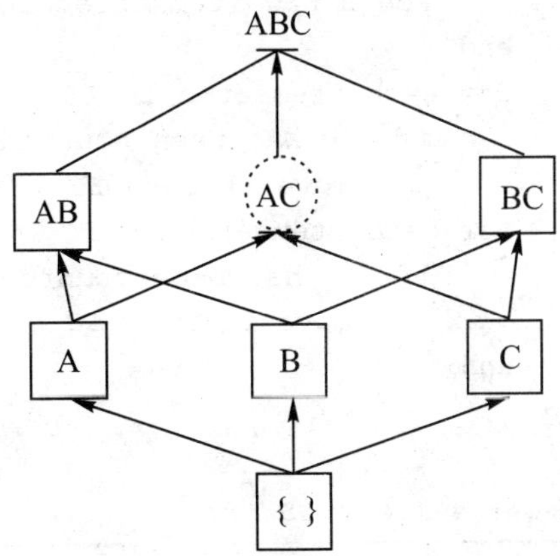

Counters: A = 2, B = 2, C = 1, AB = 1, AC = 0, BC = 1. AC and BC are counted all the way through. We do not count ABC because one of its subsets is a circle. There are no dashed itemsets left so the algorithm is done.

Operations:

1. add new itemsets
2. maintain a counter for every itemset
3. manage itemset states from dashed to solid and from circle to square
4. when itemsets become large determine which new itemsets should be added because they could potentially be large.

Pseudocode Algorithm:

```
SS = Æ ;  // solid square (frequent)
SC = Æ ;  // solid circle (infrequent)
DS = Æ ;  // dashed square (suspected frequent)
DC = { all 1-itemsets }; // dashed circle (suspected infrequent)
while (DS != 0) or (DC != 0) do begin
    read M transactions from database into T
forall transactions t ∈ T do begin
    //increment the respective counters of the itemsets marked with dash
    for each itemset c in DS or DC do begin
        if ( c ∈ t ) then
            c.counter++ ;
    for each itemset c in DC
        if ( c.counter ≥ threshold ) then
            move c from DC to DS ;
            if(any immediate superset sc of c has all of its subsets in SS
or DS)
```

```
then
        add a new itemset sc in DC ;
    end
    for each itemset c in DS
        if ( c has been counted through all transactions ) then
            move it into SS ;
    for each itemset c in DC
        if ( c has been counted through all transactions ) then
            move it into SC ;
    end
end

Answer = { c ∈ SS } ;
```

3.10 MINING FREQUENT PATTERNS WITHOUT CANDIDATE GENERATION (FP_GROWTH)

FP-Tree:

It consists of one root labeled as null, a set of item prefix subtrees as the children of the root, and a frequent .item header table. Each node in the item prefix subtree consists of three fields: item-name, count and node link where— item-name registers which item the node represents; count registers the number of transactions represented by the portion of path reaching this node, node link links to the next node in the FP- tree. Each item in the header table consists of two fields—item name and head of node link, which points to the first node in the FP-tree carrying the item name.

Algorithm 1(FP-tree construction)

(1) Scan the transaction database DB once. Collect the set of frequent items F and their supports. Sort F in support descending order as L, the list of frequent items.

(2) Create the root of an FP-tree, T, and label it as null. For each transaction in the database do the following.

Select and sort the frequent item in each transaction according to the order of L

Let the sorted frequent item list be [p|P], where p is the first element and P is the remaining list. Call insert for each item insert ([p|P], T).

```
-insert function.
insert([p|P],T)
{
// check if T has a child N where N.item-name=p.item-name then increment N
count by 1
//else create a new node with count 1,
Its parent linked to T,and its node-link be linked to nodes with the same
item-name via node-link structure //call insert till P is non empty.
}
```

Based on the FP-Tree characteristic, we can develop an efficient mining method for mining the complete set of frequent patterns. FP-growth method transforms the problem of finding long frequent pattern to looking for shorter ones recursively and then concatenating the suffix. It uses the least frequent items as a suffix, offering good selectivity.

The mining of the FP-tree proceeds as follows. Start from each frequent length-1 pattern; construct its conditional pattern base (a sub database which consists of the set of prefix paths in the FP-tree co-occurring with the suffix pattern), then construct its FP-tree and perform mining recursively on such a tree.

Algorithm 2

```
Input: constructed FP-tree
Output: complete set of frequent patterns
Method: Call FP-growth (FP-tree, null).
procedure FP-growth (Tree, α)
procedure FP-growth (Tree, α)
{
if Tree contains a single path P
then for each combination do generate pattern β α with support = minimum
support of nodes in β.
Else
For each header ai in the header of Tree do
{
Generate pattern β = ai α with
support = ai.support;
Construct α.s conditional pattern base and
then β.s conditional FP-tree Tree β
If Tree? <> null
Then call FP-growth (Tree β, β)}
}
```

Opinion:

- Compare to Apriori-like algorithm, it is difficult to implement FP-Tree on actual coding because of the tree structure.
- In case of large database, it is a good candidate to use for short and long patterns because FP-growth scales much better than Apriori. It becomes very obvious when the .support threshold. goes down.
- FP-tree is constructed the way that the higher frequent items are closer to the root (upper portion). Therefore, from a searching or scanning point of view, it is very easy to select (mining) the items with high threshold by dropping the lower portion of the tree.

3.11 IMPROVING THE EFFICIENCY OF ASSOCIATION RULE MINING

The efficiency of association rule mining can be increased by following methods:

A. Sampling

Sampling is a powerful data reduction technique that has been applied to a variety of data mining algorithms for reducing computational overhead. In the context of association rules, sampling can be utilized to gather quick preliminary rules. This may help the user to direct the data mining process by refining the criterion for "interesting" rules. Sampling can speed up the mining process by more than an order of magnitude by reducing I/O costs and drastically shrinking the number of transaction to be considered. The validity of the sample is determined by two characteristics the size of the sample and the quality of the sample.

The quality, in the context of statistical sampling techniques, refers to whether the sample captures the characteristics of the database. The highest quality sample would be an exact miniature of the database; it would preserve the distributions of individual variables and the relationships among variables. The quality of the sample for association rule mining can be improved by considering factors like transaction length and transaction frequency.

A number of studies were conducted to propose efficient methods for mining association rules by reducing either the CPU computation time or the disk access overhead. Some studies considered the usage of sampling techniques for reducing the processing overhead. Most of the prior works on sampling have concentrated on speeding up the phase by running a frequent itemset mining algorithm only on a small sample of the database. Chiefly, researchers have evaluated the viability of using sampling to reduce the dataset size. While such methods have shown quite a lot of promise it has been observed by several researchers that it is often very difficult to quantify, apriori, the quality of the results obtained for a given sample size, necessitating novel and more effective sampling-based association rule mining algorithms to foster better mining results.

B. Reducing the number of passes

The disadvantage of Apriori algorithm made the researchers to think about new techniques to mine frequent patterns. The two main negative sides are the possible need of generating a huge number of candidates if the number of frequent 1-itemsets is high or if the size of the frequent pattern is big, the database has to be scanned twice repeatedly to match the candidates and determine the support. What if we find a way to mine the frequent patterns without candidate generation? This would be a big improvement over Apriori. That is what the frequent pattern growth (FP-growth) algorithm does.

Wang et al presented PRICES, an efficient algorithm for mining association rules. Their approach reduces large itemset generation time, which dictates most of the time in generating candidates by scanning the database only once and using logical operations in the process.

Another algorithm called Matrix Algorithm generates a matrix which entries 1 or 0 by passing over the cruel database only once, and then the frequent candidate sets are obtained from the resulting matrix. Association rules are then mined from the frequent candidate sets.

C. Hash-based itemset counting

A hash technique is very efficient in generating the candidate item sets, in particular for the large two-itemsets, thus greatly improving the performance bottleneck of the entire process. Soo et al proposed Direct Hashing and Pruning [DHP] algorithm, an effective hash based technique for mining the association rules. This algorithm employs effective pruning techniques to progressively reduce the transaction database size. DHP utilizes a hashing technique to filter the ineffective candidate frequent 2 itemsets. DHP also avoids database scans in some passes as to reduce the disk I/O cost involved. Another novel hash-based approach for mining frequent item-sets over data streams was developed by En *et al*. The algorithm compresses the information of all itemsets into a structure with a fixed hash-based technique. This approach skillfully summarizes the information of the whole data stream by using a hash table to estimate the support counts of the non-frequent itemsets, and keeps only the frequent itemsets for speeding up the mining process.

Another algorithm Inverted Hashing and Pruning (IHP) proposed by John et al. for is developed for mining association rules between words in text databases. The characteristics of text databases are quite different from those of retail transaction databases, and existing mining algorithms cannot handle text databases efficiently, because of the large number of itemsets (*i.e.*, words) that need to be counted. Two well-known mining algorithms, the Apriori algorithm and Direct Hashing and Pruning (DHP) algorithm, are evaluated in the context of mining text databases, and are compared with the proposed IHP algorithm. It has been shown that the IHP algorithm has better performance for large text databases.

D. Transaction Reduction

Transaction reduction is another way that helps in mining association rules effectively. It relies on a concept that a transaction that does not contain any frequent k-itemset is useless in subsequent scans. AprioriTid algorithm is another way of improving the performance of Association Rules. This algorithm is used to construct the frequent itemset. The main idea of all these algorithm is according to the theory that the subset of a frequent itemset is a frequent itemset and the superset of an infrequent itemset is an infrequent itemset. They scan the database repeatedly to mining the association rules. There is another feature for algorithm AprioriTID, the support of the candidate frequent itemsets are calculated only at the first time it scanned the database D and also generated candidate transaction database D′ which only includes the candidate frequent itemsets. Then the latter mining are based on the database D′, It reduce the time of I/O operation because D′ is smaller than D, so, it enhance the efficiency of the algorithm. Another approach called MTR-FMA (modified transaction reduction based frequent itemset mining algorithm) developed by Thevar et al maintains its performance even at relative low supports.

E. Partitioning

Various approaches to generate large item sets have been proposed based on portioning of the set of transactions. In this case, D is divided into *p* partitions D_1, D_2....D_p. Partitioning may improve the performance of finding large item sets in several ways. By using partitioning, parallel and/or distributed algorithms can be easily created, where each partition could be handled by a separate machine.

Cheung et al presented an algorithm called FDM. FDM is a parallelization of Apriori to shared nothing machines, each with its own partition of the database. At every level and on each machine, the database scan is performed independently on the local partition. Distributed Pruning is then done.

FPM (Fast Parallel Mining) for Association rule mining has been proposed. It adopts Count Distribution approach and has incorporated two powerful candidate pruning techniques. It has a simple communication scheme which performs only one round of message exchange in each iteration.

Parthasarathy et al have presented an exceilent survey on parallel association rule mining with shared memory architecture covering most of challenges and approaches adopted for parallel data mining.

F. Adding extra constraints

Another type of association rule mining involves in retrieving patterns by adding extra constraints on the structure of patterns. Techniques applicable to constraint-driven pattern discovery can be classified into the following groups:

- Post-processing (filtering out patterns that do not satisfy user-specified constraints after the actual discovery process;
- Pattern filtering (integration of pattern constraints into the actual mining process in order to generate only patterns satisfy pattern constraints.

Dataset filtering (restricting the source data set to objects that can possibly contain patterns that satisfy pattern constraints)

SUMMARY

- Association rule mining discovers relationships among attributes in databases, producing if-then statements concerning attribute-values.
- Association rules capture the co-occurrence of items or events in large volumes of customer transaction data.
- Classification refers to assigning cases into categories based on a predictable attribute.
- **Clustering** is used to identify natural groupings of cases based on a set of attributes.
- Association is interaction and data communication to provide useful information.
- Regression tasks can solve many business problems.
- **Deviation analysis** is for finding those rare cases that behave very differently from others.
- Apriori is designed to operate on databases containing transactions, collections of items bought by customers, or details of a website frequentation.
- Apriori uses breadth-first search and a tree structure to count candidate item sets efficiently.
- The APRIORI algorithm uses the downward closure property, to prune unnecessary branches for further consideration.
- The DHP algorithm employed a hash mechanism to filter out unuseful itemsets.
- A dynamic itemset counting technique was proposed in which the database is partitioned into blocks marked by start points.
- An itemset lattice contains all of the possible itemsets for a transaction database.

REVIEW QUESTIONS

1. Write a short note on Association Rule Mining.
2. Compare the APRIORI and APRIORI-Tid algorithms.
3. How can we improve the efficiency of Association Mining Rule algorithms?
4. Explain DIP with the help of an illustrative example.
5. How the efficiency of DIC can be improved?
6. Compare the performance of Association Mining Rule algorithms.

■■■

CHAPTER 4

Classification and Prediction

Two common data mining techniques for finding hidden patterns in data are clustering and classification analyses. Although classification and clustering are often mentioned in the same breath, they are different analytical approaches. Imaging a database of customer records, where each record represents a customer's attributes. These can include identifiers such as name and address, demographic information such as gender and age, and financial attributes such as income and revenue spent.

Clustering is an automated process to group related records together. Related records are grouped together on the basis of having similar values for attributes. This approach of segmenting the database via clustering analysis is often used as an exploratory technique because it is not necessary for the end-user/analyst to specify ahead of time how records should be related together. In fact, the objective of the analysis is often to discover segments or clusters, and then examine the attributes and values that define the clusters or segments. As such, interesting and surprising ways of grouping customers together can become apparent, and this in turn can be used to drive marketing and promotion strategies to target specific types of customers.

There are a variety of algorithms used for clustering, but they all share the property of iteratively assigning records to a cluster, calculating a measure (usually similarity, and/or distinctiveness), and re-assigning records to clusters until the calculated measures don't change much indicating that the process has converged to stable segments. Records within a cluster are more similar to each other, and more different from records that are in other clusters. Depending on the particular implementation, there are a variety of measures of similarity that are used (*e.g.,* based on spatial distance, based on statistical variability, or even adaptations of Condorcet values used in voting schemes), but the overall goal is for the approach to converge to groups of related records.

Classification is a different technique than clustering. Classification is similar to clustering in that it also segments customer records into distinct segments called classes. But unlike clustering, a classification analysis requires that the end-user/analyst know ahead of time how classes are defined. For example, classes can be defined to represent the likelihood that a customer defaults on a loan (Yes/No). It is necessary that each record in the dataset used to build the classifier already have a value for the attribute used to define classes. Because each record has a value for the attribute used to define the classes, and because the end-user decides on the attribute to use, classification is much less exploratory than clustering. The objective of a classifier is not to explore the data to discover interesting segments, but rather to decide how new records should be classified *i.e.,* is this new customer likely to default on the loan?

Classification routines in data mining also use a variety of algorithms and the particular algorithm used can affect the way records are classified. A common approach for classifiers is to use decision trees to partition and segment records. New records can be classified by traversing the tree from the root through branches and nodes, to a leaf representing a class. The path a record takes through a decision tree can then be represented as a rule.

For example, "Income < $30,000 and age < 25, and debt = High, then Default Class = Yes). But due to the sequential nature of the way a decision tree splits records (*i.e.* the most discriminative attribute-values [*e.g.*, Income] appear early in the tree) can result in a decision tree being overly sensitive to initial splits. Therefore, in evaluating the goodness of fit of a tree, it is important to examine the error rate for each leaf node (proportion of records incorrectly classified). A nice property of decision tree classifiers is that because paths can be expressed as rules, then it becomes possible to use measures for evaluating the usefulness of rules such as Support, Confidence and Lift to also evaluate the usefulness of the tree.

Data prediction is a two step process, similar to that of data classification. However, for prediction, instead of using the term "class label attribute", the attribute can be referred to simply as the predicted attribute. Prediction and classification also differ in the methods that are used to build their respective models. As with classification, the training set used to build a predictor should not be used to assess its accuracy. An independent test set should be used instead. The accuracy of a predictor is estimated by computing an error based on the difference between the predicted value and the actual known value of *y* for each of the test tuples, X.

4.1 CLASSIFICATION AND PREDICTION—INTRODUCTION

The models used in data mining are grouped under two categories; predictive and descriptive. The goal of the predictive models is to construct a model by using the results of the known data and is to predict the results of unknown data sets by using the constructed model. For instance a bank might have the necessary data about the loans given in the previous terms. In this data, independent variables are the characteristics of the loan granted clients and the dependent variable is whether the loan is paid back or not. The model constructed by this data is used in the prediction of whether the loan will be paid back by client in the next loan applications.

However in the descriptive models, patterns in the current data which will guide the decision making process are defined. The determination of the similarities between the families with children which have two or more cars and whose income is in the interval X-Y, and the families with no children whose income is less than X-Y interval is an example of these descriptive models. Data mining models can be divided into three groups according to their functions:

1. Classification and Regression;
2. Clustering;
3. Association Rules.

Classification and regression models are predictive and clustering and association rules models are descriptive. Classification and regression are two data analyzing methods which determine important data classes or may construct models which can predict future data trends. While the classification predicts the categorical values, the regression is used in the prediction of values showing continuity. For instance while the classification model is constructed to categorize whether the bank loan applications are safe or risky, the regression model may be constructed to predict the spending of clients buying computer products whose income and occupation are given.

In the classification and regression models the following techniques are mainly used:

1. Decision Trees;
2. Artificial Neural Networks;
3. Genetics Algorithm;
4. K-Nearest Neighbour;
5. Memory Based Reasoning;
6. Navie-bayes.

Classification is a data mining function that assigns items in a collection to target categories or classes. The goal of classification is to accurately predict the target class for each case in the data. For example, a classification model could be used to identify loan applicants as low, medium, or high credit risks.

A classification task begins with a data set in which the class assignments are known. For example, a classification model that predicts credit risk could be developed based on observed data for many loan applicants over a period of time. In addition to the historical credit rating, the data might track employment history, home ownership or rental, years of residence, number and type of investments, and so on. Credit rating would be the target, the other attributes would be the predictors, and the data for each customer would constitute a case.

Classifications are discrete and do not imply order. Continuous, floating-point values would indicate a numerical, rather than a categorical, target. A predictive model with a numerical target uses a regression algorithm, not a classification algorithm.

The simplest type of classification problem is binary classification. In binary classification, the target attribute has only two possible values: for example, high credit rating or low credit rating. Multiclass targets have more than two values: for example, low, medium, high, or unknown credit rating.

In the model build (training) process, a classification algorithm finds relationships between the values of the predictors and the values of the target. Different classification algorithms use different techniques for finding relationships. These relationships are summarized in a model, which can then be applied to a different data set in which the class assignments are unknown.

Classification models are tested by comparing the predicted values to known target values in a set of test data. The historical data for a classification project is typically divided into two data sets: one for building the model; the other for testing the model. Scoring a classification model results in class assignments and probabilities for each case. For example, a model that classifies customers as low, medium, or high value would also predict the probability of each classification for each customer.

Classification has many applications in customer segmentation, business modeling, marketing, credit analysis, and biomedical and drug response modelling.

4.2 ISSUES REGARDING CLASSIFICATION AND PREDICTION

Various issues regarding Classification and Prediction are as follows:

A. Preparing the Data for Classification and Prediction

The following preprocessing steps may be applied to the data to help improve the accuracy, efficiency, and scalability of the classification or prediction process.

- **Data cleaning:** This refers to the preprocessing of data in order to remove or reduce noise and the treatment of missing values. This step can help reduce confusion during learning.

- **Relevance analysis:** Many of the attributes in the data may be redundant. A database may also contain irrelevant attributes. Hence, relevance analysis, in the form of correlation analysis and attribute subset selection, can be used to detect attributes that do not contribute to the classification or prediction task.

  Ideally, the time spent on relevance analysis, when added to the time spent on learning from the resulting "reduced" attribute subset, should be less than the time that would have been spent on learning from the original set of attributes. Hence, such analysis can help improve classification efficiency and scalability.
- **Data transformation and reduction:** Normalization involves scaling all values for a given attribute so that they fall within a small specified range, such as 0:0 to 1:0. The data can also be transformed by generalizing it to higher-level concepts. Concept hierarchies may be used for this purpose.

  Data can also be reduced by applying many other methods, ranging from wavelet transformation and principle components analysis to discretization techniques, such as binning, histogram analysis, and clustering.

B. Comparing Classification and Prediction Methods

Classification and prediction methods can be compared and evaluated according to the following criteria:

- **Accuracy:** The accuracy of a classifier refers to the ability of a given classifier to correctly predict the class label of new or previously unseen data. Similarly, the accuracy of a predictor refers to how well a given predictor can guess the value of the predicted attribute for new or previously unseen data. Estimation techniques are cross-validation and bootstrapping. Because the accuracy computed is only an estimate of how well the classifier or predictor will do on new data tuples, confidence limits can be computed to help gauge this estimate.
- **Speed:** This refers to the computational costs involved in generating and using the given classifier or predictor.
- **Robustness:** This is the ability of the classifier or predictor to make correct predictions given noisy data or data with missing values.
- **Scalability:** This refers to the ability to construct the classifier or predictor efficiently given large amounts of data.
- **Interpretability:** This refers to the level of understanding and insight that is provided by the classifier or predictor. Interpretability is subjective and therefore more difficult to assess.

4.3 CLASSIFICATION METHODS AND ALGORITHMS

Oracle Data Mining provides the following algorithms for classification:

- **Decision Tree:** Decision trees automatically generate rules, which are conditional statements that reveal the logic used to build the tree.
- **Naive Bayes:** Naive Bayes uses Bayes' Theorem, a formula that calculates a probability by counting the frequency of values and combinations of values in the historical data.
- **Generalized Linear Models (GLM):** GLM is a popular statistical technique for linear modeling. Oracle Data Mining implements GLM for binary classification and for

regression. GLM provides extensive coefficient statistics and model statistics, as well as row diagnostics. GLM also supports confidence bounds.

- **Support Vector Machine:** Support Vector Machine (SVM) is a powerful, state-of-the-art algorithm based on linear and nonlinear regression. Oracle Data Mining implements SVM for binary and multiclass classification.

4.4 DECISION TREE

Decision trees are powerful and popular tools for classification and prediction. The attractiveness of decision trees is due to the fact that, in contrast to neural networks, decision trees represent *rules*. Rules can readily be expressed so that humans can understand them or even directly used in a database access language like SQL so that records falling into a particular category may be retrieved.

In some applications, the accuracy of a classification or prediction is the only thing that matters. In such situations we do not necessarily care how or why the model works. In other situations, the ability to explain the reason for a decision is crucial. In marketing one has describe the customer segments to marketing professionals, so that they can utilize this knowledge in launching a successful marketing campaign. This domain expert must recognize and approve this discovered knowledge, and for this we need good descriptions. There are a variety of algorithms for building decision trees that share the desirable quality of interpretability. A well known and frequently used over the years is C4.5.

Decision tree is a classifier in the form of a tree structure, where each node is either:

- a *leaf node*: indicates the value of the target attribute (class) of examples, or
- a *decision node*: specifies some test to be carried out on a single attribute-value, with one branch and sub-tree for each possible outcome of the test.

A decision tree can be used to classify an example by starting at the root of the tree and moving through it until a leaf node, which provides the classification of the instance.

Decision tree induction is a typical inductive approach to learn knowledge on classification. The key requirements to do mining with decision trees are:

- *Attribute-value description*: object or case must be expressible in terms of a fixed collection of properties or attributes. This means that we need to discretize continuous attributes, or this must have been provided in the algorithm.
- *Predefined classes (target attribute values)*: The categories to which examples are to be assigned must have been established beforehand (supervised data).
- *Discrete classes*: A case does or does not belong to a particular class, and there must be more cases than classes.
- *Sufficient data*: Usually hundreds or even thousands of training cases.

Decision Tree Rules

The decision tree method is a popular and relatively simple supervised classification method that involves each node of the tree specifying a test of some attribute and each branch from the node corresponding to one of the values of the attribute. Each path from the root to leaf of the decision tree therefore consists of attribute tests, finally reaching a leaf that describes the class. The popularity of decision trees is partly due to the ease of understanding the rules that the nodes specify. One could even use the rules specified by decision tree to retrieve data from a relational database satisfying the rules using SQL.

There are a number of advantages in converting a decision tree to rules. Decision rules make it easier to make pruning decisions since it is easier to see the context of each rule. Also, converting to rules removes the distinction between attribute tests that occur near the roof of the tree and they are easier for people to understand. IF-THEN rules may be derived based on the various paths from the root to the leaf nodes. Although the simple approach will lead to as many rules as the leaf nodes, rules can often be combined to produce a smaller set of rules. For example:

If Gender = "Male" then Class = B

If Gender = "Female" and Married = "Yes" then Class = C, else Class = A

Once all the rules have been generated, it may be possible to simplify the rules. Rules with only one antecedent (*e.g.,* if Gender = "Male" then Class = B) cannot be further simplified, so we only consider those with two or more antecedents. It may be possible to eliminate unnecessary rule antecedents that have no effect on the conclusion reached by the rule. Some rules may be unnecessary and these may be removed. In some cases a number of rules that lead to the same class may be combined.

Constructing decision trees

Most algorithms that have been developed for learning decision trees are variations on a core algorithm that employs a top-down, greedy search through the space of possible decision trees. Decision tree programs construct a decision tree T from a set of training cases.

J. Ross Quinlan originally developed ID3 at the University of Sydney. He first presented ID3 in 1975 in a book, *Machine Learning*, vol. 1, no. 1. ID3 is based on the Concept Learning System (CLS) algorithm.

ID3 Decision Tree Algorithm

```
function ID3
Input: (R: a set of non-target attributes,
   C: the target attribute,
   S: a training set) returns a decision tree;
begin
 If S is empty, return a single node with
   value Failure;
 If S consists of records all with the same
   value for the target attribute,
   return a single leaf node with that value;
 If R is empty, then return a single node
    with the value of the most frequent of the
    values of the target attribute that are
    found in records of S; [in that case
    there may be be errors, examples
    that will be improperly classified];
 Let A be the attribute with largest
 Gain(A,S) among attributes in R;
 Let {aj| j=1,2, .., m} be the values of
    attribute A;
```

```
Let {Sj| j=1,2, .., m} be the subsets of
S consisting respectively of records
   with value aj for A;
Return a tree with root labeled A and arcs
   labeled a1, a2, .., am going respectively
   to the trees (ID3(R-{A}, C, S1), ID3(R-{A}, C, S2),
   .....,ID3(R-{A}, C, Sm);
Recursively apply ID3 to subsets {Sj| j=1,2, .., m}
   until they are empty
end
```

ID3 searches through the attributes of the training instances and extracts the attribute that best separates the given examples. If the attribute perfectly classifies the training sets then ID3 stops; otherwise it recursively operates on the m (where m = number of possible values of an attribute) partitioned subsets to get their "best" attribute. The algorithm uses a greedy search, that is, it picks the best attribute and never looks back to reconsider earlier choices. Note that ID3 may misclassify data.

The central focus of the decision tree growing algorithm is selecting which attribute to test at each node in the tree. For the selection of the attribute with the most inhomogeneous class distribution the algorithm uses the concept of entropy, which is explained next

The estimation criterion in the decision tree algorithm is the selection of an attribute to test at each decision node in the tree. The goal is to select the attribute that is most useful for classifying examples. A good quantitative measure of the worth of an attribute is a statistical property called *information gain* that measures how well a given attribute separates the training examples according to their target classification. This measure is used to select among the candidate attributes at each step while growing the tree.

Entropy

In order to define information gain precisely, we need to define a measure commonly used in information theory, called entropy that characterizes the impurity of an arbitrary collection of examples. Given a set S, containing only positive and negative examples of some target concept (a2 class problem), the entropy of set S relative to this simple, binary classification is defined as:

$$\text{Entropy(S)} = - p_p \log_2 p_p - p_n \log_2 p_n$$

where p_p is the proportion of positive examples in S and p_n is the proportion of negative examples in S. In all calculations involving entropy we define 0 log 0 to be 0.

To illustrate, suppose S is a collection of 25 examples, including 15 positive and 10 negative examples [15+, 10–]. Then the entropy of S relative to this classification is

$$\text{Entropy(S)} = - (15/25) \log_2 (15/25) - (10/25) \log_2 (10/25) = 0.970$$

Notice that the entropy is 0 if all members of S belong to the same class. For example, if all members are positive (p_p= 1), then p_n is 0, and Entropy(S) = $-1' \log_2(1) - 0' \log_2 0 = -1' 0 - 0' \log_2 0$ = 0. Note the entropy is 1 (at its maximum!) when the collection contains an equal number of positive and negative examples. If the collection contains unequal numbers of positive and negative examples, the entropy is between 0 and 1.

One interpretation of entropy from information theory is that it specifies the minimum number of bits of information needed to encode the classification of an arbitrary member of S (*i.e.*, a member of S drawn at random with uniform probability). For example, if p_p is 1, the receiver knows the drawn example will be positive, so no message need be sent, and the entropy is 0. On the other hand, if p_p is 0.5, one bit is required to indicate whether the drawn example is positive or negative. If p_p is 0.8, then a collection of messages can be encoded using on average less than 1 bit per message by assigning shorter codes to collections of positive examples and longer codes to less likely negative examples.

Thus far we have discussed entropy in the special case where the target classification is binary. If the target attribute takes on *c* different values, then the entropy of S relative to this *c*-wise classification is defined as

$$\text{Entropy(S)} = \sum_{i=1}^{0} - p_i \log_2 p_i$$

where p_i is the proportion of S belonging to class *i*. Note the logarithm is still base 2 because entropy is a measure of the expected encoding length measured in bits. Note also that if the target attribute can take on *c* possible values, the maximum possible entropy is $\log_2 c$.

Information gain measures the expected reduction in entropy

Given entropy as a measure of the impurity in a collection of training examples, we can now define a measure of the effectiveness of an attribute in classifying the training data. The measure we will use, called *information gain*, is simply the expected reduction in entropy caused by partitioning the examples according to this attribute. More precisely, the information gain, *Gain* (*S, A*) of an attribute *A*, relative to a collection of examples S, is defined as

$$\text{Gain (S,A)} = \text{Entropy (S)} - \sum_{v \in \text{Values (A)}} \frac{|S_v|}{|S|} \text{Entropy } (S_v)$$

where *Values* (A) is the set of all possible values for attribute A, and S_v is the subset of S for which attribute *A* has value *v* (*i.e.*, $S_v = \{s\} \; S \mid A(s) = v\}$). Note the first term in the equation for *Gain* is just the entropy of the original collection S and the second term is the expected value of the entropy after S is partitioned using attribute A. The expected entropy described by this second term is simply the sum of the entropies of each subset S_v, weighted by the fraction of examples $|S_v|/|S|$ that belong to S_v. *Gain* (S, A) is therefore the expected reduction in entropy caused by knowing the value of attribute *A*. Put another way, *Gain* (S, A) is the information provided about the target attribute value, given the value of some other attribute A. The value of *Gain* (S, A) is the number of bits saved when encoding the target value of an arbitrary member of S, by knowing the value of attribute A.

The process of selecting a new attribute and partitioning the training examples is now repeated for each non-terminal descendant node, this time using only the training examples associated with that node. Attributes that have been incorporated higher in the tree are excluded, so that any given attribute can appear at most once along any path through the tree. This process continues for each new leaf node until either of two conditions is met:

1. every attribute has already been included along this path through the tree, or
2. the training examples associated with this leaf node all have the same target attribute value (*i.e.*, their entropy is zero).

Issues in data mining with decision trees

Practical issues in learning decision trees include determining how deeply to grow the decision tree, handling continuous attributes, choosing an appropriate attribute selection measure, handling training data with missing attribute values, handing attributes with differing costs, and improving computational efficiency. Below we discuss each of these issues and extensions to the basic ID3 algorithm that address them.

Incorporating Continuous-Valued Attributes

The initial definition of ID3 is restricted to attributes that take on a discrete set of values. First, the target attribute whose value is predicted by the learned tree must be discrete valued. Second, the attributes tested in the decision nodes of the tree must also be discrete valued. This second restriction can easily be removed so that continuous-valued decision attributes can be incorporated into the learned tree. This can be accomplished by dynamically defining new discrete-valued attributes that partition the continuous attribute value into a discrete set of intervals. In particular, for an attribute *A* that is continuous-valued, the algorithm can dynamically create a new Boolean attribute A_c that is true if $A < c$ and false otherwise. The only question is how to select the best value for the threshold *c*. Clearly, we would like to pick a threshold, *c*, that produces the greatest information gain. By sorting the examples according to the continuous attribute A, then identifying adjacent examples that differ in their target classification, we can generate a set of candidate thresholds midway between the corresponding values of A. It can be shown that the value of *c* that maximizes information gain must always lie at such a boundary. These candidate thresholds can then be evaluated by computing the information gain associated with each. The information gain can then be computed for each of the candidate attributes, and the best can be selected. This dynamically created Boolean attribute can then compete with the other discrete-valued candidate attributes available for growing the decision tree.

Handling Training Examples with Missing Attribute Values

In certain cases, the available data may be missing values for some attributes. For example, in a medical domain in which we wish to predict patient outcome based on various laboratory tests, it may be that the lab test Blood-Test-Result is available only for a subset of the patients. In such cases, it is common to estimate the missing attribute value based on other examples for which this attribute has a known value.

Consider the situation in which *Gain* (S, A) is to be calculated at node *n* in the decision tree to evaluate whether the attribute A is the best attribute to test at this decision node. Suppose that $<x, c(x)>$ is one of the training examples in S and that the value A(*x*) is unknown, where *c*(*x*) is the class label of *x*.

One strategy for dealing with the missing attribute value is to assign it the value that is most common among training examples at node n. Alternatively, we might assign it the most common value among examples at node n that have the classification *c*(*x*). The elaborated training example using this estimated value for A(*x*) can then be used directly by the existing decision tree learning algorithm.

A second, more complex procedure is to assign a probability to each of the possible values of A rather than simply assigning the most common value to A(*x*). These probabilities can be estimated again based on the observed frequencies of the various values for A among the examples at node *n*. For example, given a Boolean attribute A, if node *n* contains six known examples with A = 1 and four with A = 0, then we would say the probability that A(*x*) = 1 is 0.6, and the probability that A(*x*) = 0 is 0.4. A fractional 0.6 of instance *x* is now distributed down the branch for A = 1 and a fractional 0.4 of *x* down the other tree branch. These fractional examples are used for the purpose of computing information *Gain* and can be further subdivided at subsequent branches of the tree if a second missing attribute value must be tested. This same fractioning of examples can also be applied after learning, to classify new instances whose attribute values are unknown. In this case, the classification of the new instance is simply the most probable classification, computed by summing the weights of the instance fragments classified in different ways at the leaf nodes of the tree. This method for handling missing attribute values is used in C4.5.

Strengths and Weakness of Decision Tree Methods

The strengths of decision tree methods are:

- Decision trees are able to generate understandable rules.
- Decision trees perform classification without requiring much computation.
- Decision trees are able to handle both continuous and categorical variables.
- Decision trees provide a clear indication of which fields are most important for prediction or classification.

The weaknesses of decision tree methods

- Decision trees are less appropriate for estimation tasks where the goal is to predict the value of a continuous attribute.
- Decision trees are prone to errors in classification problems with many class and relatively small number of training examples.
- Decision tree can be computationally expensive to train. The process of growing a decision tree is computationally expensive. At each node, each candidate splitting field must be sorted before its best split can be found. In some algorithms, combinations of fields are used and a search must be made for optimal combining weights. Pruning algorithms can also be expensive since many candidate sub-trees must be formed and compared.
- Decision trees do not treat well non-rectangular regions. Most decision-tree algorithms only examine a single field at a time. This leads to rectangular classification boxes that may not correspond well with the actual distribution of records in the decision space.

Decision tree learning, used in statistics, data mining and machine learning, uses a decision tree as a predictive model which maps observations about an item to conclusions about the item's target value. More descriptive names for such tree models are **classification trees** or **regression trees**. In these tree structures, leaves represent classifications and branches represent conjunctions of features that lead to those classifications.

In decision analysis, a decision tree can be used to visually and explicitly represent decisions and decision making. In data mining, a decision tree describes data but not decisions; rather the resulting classification tree can be an input for decision making. Decision tree learning is a method commonly used in data mining. The goal is to create a model that predicts the value of a target variable based on several input variables.

A tree can be "learned" by splitting the source set into subsets based on an attribute value test. This process is repeated on each derived subset in a recursive manner called recursive partitioning. The recursion is completed when the subset at a node all has the same value of the target variable, or when splitting no longer adds value to the predictions.

In data mining, trees can be described also as the combination of mathematical and computational techniques to aid the description, categorization and generalization of a given set of data.

Decision trees used in data mining are of two main types:

- **Classification tree** analysis is when the predicted outcome is the class to which the data belongs.

- **Regression tree** analysis is when the predicted outcome can be considered a real number (*e.g.*, the price of a house, or a patient's length of stay in a hospital).

The term **Classification And Regression Tree (CART)** analysis is an umbrella term used to refer to both of the above procedures, first introduced by Breiman et al. Trees used for regression and trees used for classification have some similarities but also some differences, such as the procedure used to determine where to split.

Some techniques use more than one decision tree for their analysis:

- A **Random Forest** classifier uses a number of decision trees, in order to improve the classification rate.
- **Boosted Trees** can be used for regression-type and classification-type problems.

There are many specific decision-tree algorithms. Notable ones include:

- ID3 algorithm
- C4.5 algorithm
- **CHi-squared Automatic Interaction Detector (CHAID)**. Performs multi-level splits when computing classification trees.
- MARS: extends decision trees to better handle numerical data

Appropriate Problems for Decision Tree Learning

It is a skilled job in AI to choose exactly the right learning representation/method for a particular learning task. As elaborated by Tom Mitchell, decision tree learning is best suited to problems with these characteristics:

- The background concepts describe the examples in terms of **attribute-value** pairs, and the values for each attribute range over finitely many fixed possibilities.
- The concept to be learned (Mitchell calls it the **target function**) has discrete values.
- Disjunctive descriptions might be required in the answer.

In addition to this, decision tree learning is robust to errors in the data. In particular, it will function well in the light of (*i*) errors in the classification instances provided (*ii*) errors in the attribute-value pairs provided and (*iii*) missing values for certain attributes for certain examples.

4.5 BAYESIAN CLASSIFICATION—NAIVE BAYES METHOD

The Naïve Bayes method is based on the work of Thomas Bayes. Bayes was a British minister and his theory was published only after his death. It is a mystery what Bayes wanted to do with such calculations. Bayesian classification is quite different from the decision tree approach. In Bayesian classification we have a hypothesis that the given data belongs to a particular class. We then calculate the probability for the hypothesis to be true. This is among the most practical approaches for certain types of problems. The approach requires only one scan of the whole data. Also, if at some stage there are additional training data then each training example can incrementally increase/decrease the probability that a hypothesis is correct.

Before we define the Bayes theorem, we will define some notation. The expression P(A) refers to the probability that event A will occur. P(A|B) stands for the probability that event A will happen, given that event B has already happened. In other words P(A|B) is the conditional probability of A based on the condition that B has already happened. For example, A and B may be probabilities of passing another course B respectively. P(A|B) then is the probability of passing A when we know that B has been passed.

Now here is the Bayes theorem:

$$P(A\backslash B) = P(B|A)\,P(A)/P(B)$$

Once might wonder where did this theorem come from. Actually it is rather easy to derive since we know the following:

$$P(A|B) = P(A \text{ and } B)/P(B)$$

and

$$P(B|A) = P(A \text{ and } B)/P(A)$$

Diving the first equation by the second gives us the Bayes' theorem. Continuing with A and B being courses, we can compute the conditional probabilities if we knew what the probability of passing both courses was, that is P(A and B), and what the probabilities of passing A and B separately were. If an event has already happened then we divide the joint probability P(A and B) with the probability of what has just happened and obtain the conditional probability.

If we consider X to be an object to be classified then Bayes' theorem may be read as giving the probability of it belonging to one of the classes C1, C2, C3 etc by calculating P(Ci|X). Once these probabilities have been computed for all the classes, we simply assign X to the class that has the highest conditional probability.

Let us now consider how probabilities P(Ci|X) may be calculated. We have P(Ci|X) = [P(X|Ci)P(Ci)]/P(X) • P(Ci|X) is the probability of the object X belonging to class Ci.

- P(X|Ci) is the probability of obtaining attribute values X if we know that it belongs to class Ci.
- P(Ci) is the probability of any object belonging to class Ci without any other information.
- P(X) is the probability of obtaining attribute values X whatever class the object belongs to.

Given the attribute values X, what probabilities in the formula can we compute? The probabilities we need to compute are P(X|Ci), P(Ci) and P(X). Actually the denominator P(X) is independent of Ci and is not required to be known since we are interested only in comparing probabilities P(Ci|X). Therefore we only need to compute P(X|Ci) and P(Ci) for each class. Comparing P(Ci) is rather easy since we count the number of instances of each class in the training data and divide each by the total number of instances. This may not be the most accurate estimation of P(Ci) but we have very little information, the training sample, and we have no other information to obtain a better estimate. This estimate will be reasonable if the training sample was large and was randomly chosen.

To compute P(X|Ci) we use a naïve approach (that is why it is called the Naïve Bayes model) by assuming that all attributes of X are independent which is often not true. Using the independence of attributes assumption and based on the training data, we compute an estimate of the probability of obtaining the data X that we have by estimating the probability of each of the attribute values by counting the frequency of those values for class Ci.

We then determine the class allocation of X by computing [P(X|Ci) P(Ci)] for each of the classes and allocating X to the class with the largest value. The beauty of the Bayesian approach is that the probability of the dependent attribute can be estimated by computing estimates of the probabilities of the independent attributes.

We should also note that it is possible to use this approach even if values of all the independent attributes are not known since we can still estimate the probabilities of the attribute values that we know. This is a significant advantage of the Bayesian approach .

A **Naive Bayes classifier** is a simple probabilistic classifier based on applying **Bayes' theorem** with strong (naive) independence assumptions. A more descriptive term for the underlying probability model would be "independent feature model".

In simple terms, a naive Bayes classifier assumes that the presence (or absence) of a particular feature of a class is unrelated to the presence (or absence) of any other feature. For example, a fruit may be considered to be an apple if it is red, round, and about 4″ in diameter. Even if these features depend on each other or upon the existence of the other features, a naive Bayes classifier considers all of these properties to independently contribute to the probability that this fruit is an apple.

Depending on the precise nature of the probability model, naive Bayes classifiers can be trained very efficiently in a supervised learning setting. In many practical applications, parameter estimation for naive Bayes models uses the method of maximum likelihood; in other words, one can work with the naive Bayes model without believing in Bayesian probability or using any Bayesian methods.

In spite of their naive design and apparently over-simplified assumptions, naive Bayes classifiers have worked quite well in many complex real-world situations. In 2004, analysis of the Bayesian classification problem has shown that there are some theoretical reasons for the apparently unreasonable efficacy of naive Bayes classifiers. Still, a comprehensive comparison with other classification methods in 2006 showed that Bayes classification is outperformed by more current approaches, such as boosted trees or random forests.

An advantage of the naive Bayes classifier is that it only requires a small amount of training data to estimate the parameters (means and variances of the variables) necessary for classification. Because independent variables are assumed, only the variances of the variables for each class need to be determined and not the entire covariance matrix.

4.6 MULTILAYER FEED-FORWARD NEURAL NETWORK

Neural networks involve long training times and are therefore more suitable for applications where this is feasible. They require a number of parameters that are typically best determined empirically, such as network topology. Neural networks have been criticized for their poor interpretability. For example, it is difficult for humans to interpret the symbolic meaning behind the learned weights and of hidden units in the network. These features initially made neural networks less desirable for data mining.

Advantages of neural networks include their high tolerance of noisy data as well as their ability to classify patterns on which they have not been trained. They can be used when you may have little knowledge of the relationship between attribute and classes. They are well suited for continuous valued inputs and outputs, unlike most decision tree algorithms. They have been successful on a wide array of real-world data, including character recognition, pathology etc. Neural network algorithms are inherently parallel. Parallelization techniques can be used to speed up the computation process. In addition, several techniques have recently been developed for the extraction of rules from trained neural networks. These factors contribute towards the usefulness of neural networks for classification & prediction in data mining. A multilayer feed forward neural network consists of an input layer, one or more hidden layers, and an output layer.

The figure 3.1 is a two-layer network. A network containing two hidden layers is called a three-layer neural network. The network is feed-forward in that none of the weights cycles back to an input unit or an output unit of a previous layer. It is fully connected iin that each unit provides input to each unit in the next forward layer.

Each output unit takes as input a weighted sum of the outputs from units in the previous layer. It applies a nonlinear activation function to the weighted input. Multilayer feed-forward neural networks are able to model the class prediction as a nonlinear combination of the inputs. From a statistical point of view, they perform nonlinear regression. Multilayer feed-forward networks, given enough hiddenunits and enough training samples, can closely approximate any function.

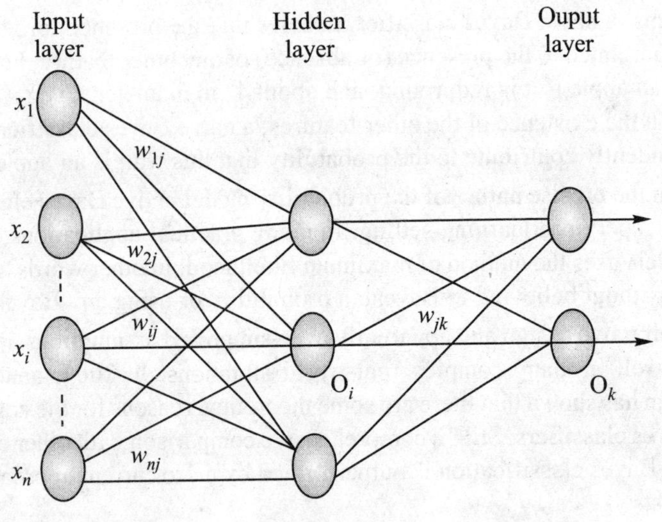

Figure 3.1. Multilayer Feed-Forward Network

4.7 CLASSIFICATION BY BACK PROPAGATION

Backpropagation is a neural network learning algorithm. A neural network is a set of input/output units in which each connection has a weight associated with it. During the learning phase, the network learns by adjusting the weights so as to be able to predict the correct class label of the input tuples. Neural Network learning is also referred to as connectionist learning due to the connection between units. The backpropagation algorithm performs learning a multilayer feed-forward neural network. It iteratively learns a set of weights for prediction of class label of tuples.

Backpropagation works on the concept of multilayer feed-forward neural networks. It learns by iteratively processing a data set of training tuples, comparing the network's prediction for each tuple with the actual known target value. The target value may be the known class label of the training tuple or a continuous value. For each training tuple, the weights are modified so as to minimize the mean squared error between the network's prediction and the actual target value. These modifications are made in the backward direction from the output layer, through each hidden layer down to the first hidden layer. Although it is not guaranteed that the weights will eventually converge and the learning process stops.

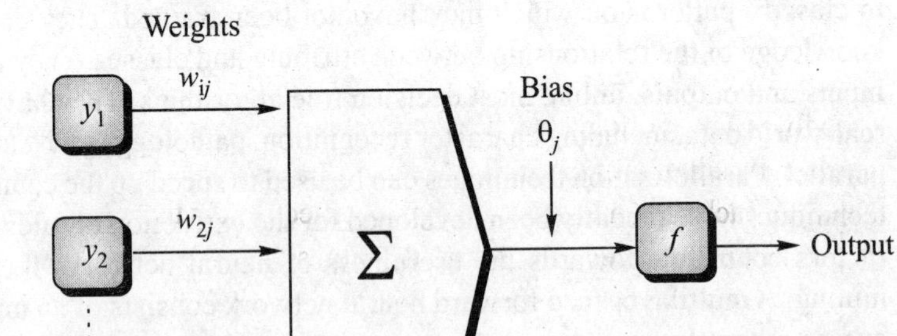

Figure 3.2. Working of BPN

4.8 CLASSIFICATION METHODS K-NEAREST NEIGHBOUR CLASSIFIERS

Nearest-neighbour classifiers are based on learning by analogy, that is, by comparing a given test tuple with training tuples that are similar to it. The training tuples are described by *n* attributes. Each tuple represents a point in an *n*-dimensional space. In this way, all of the training tuples are stored in an *n*-dimensional pattern space. When given an unknown tuple, a *k*-nearest-neighbour classifier searches the pattern space that is closest to the unknown tuple. These k training tuples are the *k* "nearest neighbours" of the unknown tuple. The unknown tuple is assigned the most common class among its *k* nearest neighbour.

Nearest-neighbour classifiers can also be used for prediction to return a real-valued prediction for a given unknown tuple. In this case, the classifier returns the average value of the real-valued labels associated with the *k* nearest neighbors of the unknown tuple.

Nearest-neighbour classifiers use distance-based comparisons that intrinsically assign equal weight to each attribute. They can thus suffer from poor accuracy when given noise or irrelevant attributes. The method has been modified to incorporate attribute weighing and pruning of noisy data tuples. The choice of a distance metric can be critical.

4.9 GENETIC ALGORITHM

Genetic algorithms are a part of soft computing techniques where the learning process starts as follows: an initial population is created consisting of randomly generated rules. Each rule can be represented by a string of bits. For example, suppose that samples in a given training set are described by two Boolean attributes, A1 and A2 and that there are two classes, C1 and C2. The rule "If A1 and not A2 then C2" can be encoded as the bit string "100", where the two leftmost bits represent attributes A1 and A2, respectively, and the rightmost bit represents the class. Similarly, the rule, "If not A1 and not A2 then C1" can be encoded as "001". If an attribute has *k* values, where $k > 2$, then *k* bits may be used to encode the attribute's values. Classes can be encoded in a similar fashion. Based on the notion of the survival of the fittest, a new population is formed as well as offsprings of these rule. Typically, the fitness of a rule is assessed by its classification accuracy on a set of training samples. Offsprings are created by applying genetic operators such as crossover and mutation. In crossover, substrings from pair of rules are swapped to form new pair of rules. In mutation, randomly selected bits in a rule's string are inverted. The process of generating new populations based on prior populations of rules continues until a population P evolves where each rule in P satisfies a prespecified fitness threshold. Genetic algorithms are easily parallelizable and have been used to evaluate the fitness of other algorithms.

4.10 TREE INDUCTION ALGORITHM

Information produced by data mining techniques can be represented in many different ways. Decision tree structures are a common way to organize classification schemes. In classifying tasks, decision trees visualize what steps are taken to arrive at a classification. Every decision tree begins with what is termed a root node, considered to be the "parent" of every other node. Each node in the tree evaluates an attribute in the data and determines which path it should follow. Typically, the decision test is based on comparing a value against some constant. Classification using a decision tree is performed by routing from the root node until arriving at a leaf node.

Figure 3.3 provided here is a canonical example in data mining, involving the decision to play or not play based on climate conditions. In this case, outlook is in the position of the root node. The degrees of the node are attribute values. In this example, the child nodes are tests of humidity and windy, leading to the leaf nodes which are the actual classifications. This example also includes the corresponding data, also referred to as instances. In our example, there are 9 "play" days and 5 "no play" days.

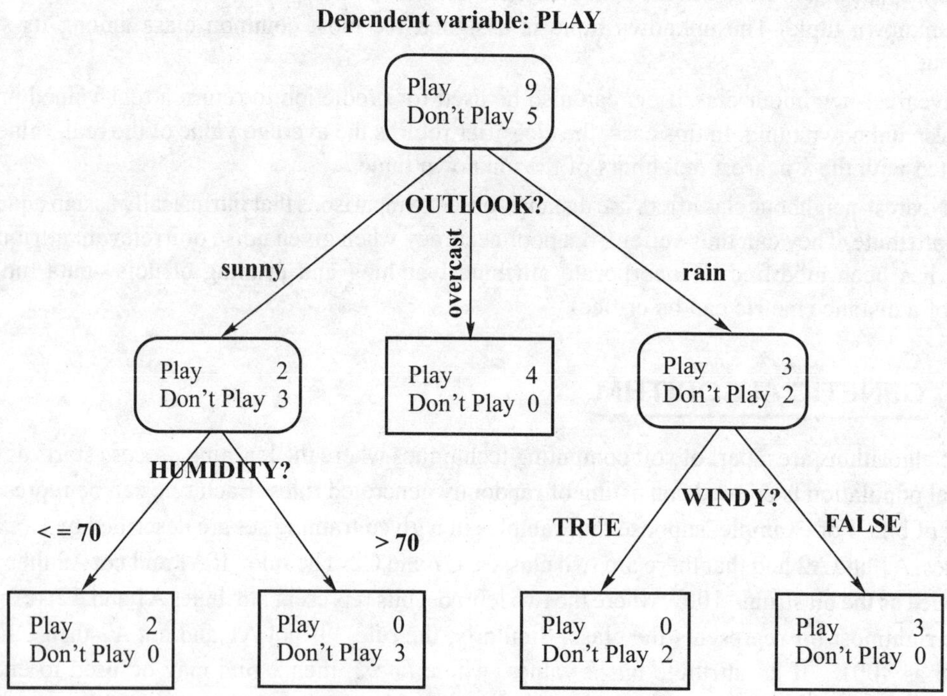

Figure 3.3. Decision Tree Induction

Decision trees can represent diverse types of data. The simplest and most familiar is numerical data. It is often desirable to organize nominal data as well. Nominal quantities are formally described by a discrete set of symbols. For example, weather can be described in either numeric or nominal fashion. We can quantify the temperature by saying that it is 11 degrees Celsius or 52 degrees Fahrenheit. We could also say that it is cold, cool, mild, warm or hot. The former is an example of numeric data, and the latter is a type of nominal data. More accurately, the example of cold, cool, mild, warm and hot is a special type of nominal data, described as ordinal data. Ordinal data has an implicit assumption of ordered relationships between the values. Continuing with the weather example, we could also have a purely nominal description like sunny, overcast and rainy. These values have no relationships or distance measures.

The type of data organized by a tree is important for understanding how the tree works at the node level. Recalling that each node is effectively a test, numeric data is often evaluated in terms of simple mathematical inequality. For example, numeric weather data could be tested by finding if it is greater than 10 degrees Fahrenheit. Nominal data is tested in Boolean fashion; in other words, whether or not it has a particular value. Figure 3.3 shows both types of tests. In the weather example, outlook is a nominal data type. The test simply asks which attribute value is represented and routes accordingly.

The humidity node reflects numeric tests, with an inequality of less than or equal to 70, or greater than 70.

Decision tree induction algorithms function recursively. First, an attribute must be selected as the root node. In order to create the most efficient (*i.e.*, smallest) tree, the root node must effectively split the data. Each split attempts to pare down a set of instances (the actual data) until they all have the same classification. The best split is the one that provides what is termed the most information gain.

Information in this context comes from the concept of entropy from information theory, as developed by Claude Shannon. Although "information" has many contexts, it has a very specific mathematical meaning relating to certainty in decision making. Ideally, each split in the decision tree should bring us closer to a classification. One way to conceptualize this is to see each step along the tree as removing randomness or entropy. Information, expressed as a mathematical quantity, reflects this. For example, consider a very simple classification problem that requires creating a decision tree to decide yes or no based on some data. This is exactly the scenario visualized in the decision tree. Each attributes values will have a certain number of yes or no classifications. If there are equal numbers of yeses and no's, then there is a great deal of entropy in that value. In this situation, information reaches a maximum. Conversely, if there are only yeses or only no's the information is also zero. The entropy is low, and the attribute value is very useful for making a decision.

4.11 SPLIT ALGORITHMS BASED ON INFORMATION THEORY

One of the techniques for selecting an attribute to split a node is based on the concept of information theory or entropy. The concept is quite simple, although often difficult to understand for many. It is based on Claude Shannon's idea that if you have uncertainty then you have information and if there is no uncertainty there is no information. For example, if a coin has a head on both sides, then the result of tossing it does not product any information but if a coin is normal with a head and a tail then the result of the toss provides information.

Essentially, information is defined as $-pi \log pi$ where pi is the probability of some event. Since the probability pi is always less than 1, $\log pi$ is always negative and $-pi \log pi$ is always positive. For those who cannot recollect their high school mathematics, we note that log of 1 is always zero whatever the base, the log of any number greater than 1 is always positive and the log of any number smaller than 1 is always negative.

Also, $\log_2 (2) = 1 \; \log_2 (2n) = n \; \log_2 (1/2) = -1 \; \log_2 (1/2n) = -n$

Information of any event that is likely to have several possible outcomes is given by

$$I = \Sigma i(-pi \log pi)$$

Consider an event that can have one of two possible values. Let the possibilities of the two values be $p1$ and $p2$. Obviously if $p1$ is 1 and $p2$ is zero, then there is no information in the outcome and $I = 0$. If $p1 = 0.5$, then the information is $I = -0.5 \log (0.5) - 0.5 \log (0.5)$

This comes out to 1.0 (using log base 2) is the maximum information that you can have for an event with two possible outcomes. This is also called entropy and is in effect a measure of the minimum number of bits required to encode the information. If we consider the case of a die (singular of dice) with six possible outcomes with equal probability, then the information is given by:

$$I = 6(-1/6) \log (1/6)) = 2.585$$

Therefore three bits are required to represent the outcome of rolling a die. Of course, if the die was loaded so that there was a 50% or a 75% chance of getting a 6, then the information content of rolling the die would be lower as given below. Note that we assume that the probability of getting any of 1 to 5 is equal (that is, equal to 10% for the 50% case and 5% for the 75% case).

$$50\%: \quad I = 5(-0.1) \log (0.1)) - 0.5 \log (0.5) = 2.16$$

$$75\%: \quad I = 5(-0.05) \log (0.05)) - 0.75 \log (0.75) = 1.39$$

Therefore we will need three bits to represent the outcome of throwing a die that has 50% probability of throwing a six but only two bits when the probability is 75%.

Information Gain

Information gain is a measure of how good an attribute is for predicting the class of each of the training data. We will select the attribute with the highest information gain as the next split attribute. Perhaps the term information gain is somewhat confusing because what we really mean is that information gain is a measure of reduction in uncertainty once the value of an attribute is known. If the uncertainty is reduced by a large amount, knowing the value of the attribute has provided a lot of information and thus we have a large information gain.

Assume there are two classes, P and N, and let the set of training data S (with a total number of objects s) contain p elements of class P and n elements of class N. The amount of information is defined as

$$I = -(n/s) \log (n/s) - (p/s) \log (p/s)$$

Obviously if $p = n$, I is equal to 1 and if $p = s$ then I = 0. Therefore if there was an attribute for which almost all the objects had the same value (for example, gender when mostpeople are male), using the attribute would lead to no information gain (that is, not reduce uncertainty) because for gender = female there will be almost no objects while gender = male would have almost all the objects even before we knew the gender. If on the other hand, an attribute divided the training sample such that gender = female resulted in objects that all belong to Class A and gender = male all belong to Class B, then uncertainty has been reduced to zero and we have a large information gain.

We define information gain for sample S using attribute A as follows:

$$\text{Gain (S, A)} = I - \Sigma(ti/s)\, Ii$$

iEvalues(A)

I is the information before the split and " iEvalues(A) (ti/s) Ii is the sum of information after the split where Ii is the information of node I and ti is the number of objects in node i. Thus the total information after the split is Σ iEvalues(A) (ti/s) Ii, which is the sum of the information of each node.

Once we have computed the information gain for every remaining attribute, the attribute with the highest information gain is selected.

4.12 SPLIT ALGORITHM BASED ON GINI INDEX

Another commonly used split approach is called the Gini index which is used in the widely used packages CART and IBM Intelligent Miner. We consider an artificial example of building a decision tree classification model to classify bank loan applications by assigning applications to one of three risk classes

Owns Home?	*Married*	*Gender*	*Employed*	*Credit Rating*	*Risk Class*
Yes	Yes	Male	Yes	A	B
Yes	Yes	Female	Yes	A	A
Yes	Yes	Female	Yes	B	C
Yes	Yes	Male	No	B	B
Yes	Yes	Female	Yes	B	C
Yes	Yes	Female	Yes	B	A
Yes	Yes	Male	No	B	B
Yes	Yes	Female	Yes	A	A
Yes	Yes	Female	Yes	A	C
Yes	Yes	Female	Yes	A	C

There are 10(s = 10) samples and three classes. The frequencies of these classes are:

A = 3

B = 3

C = 4

The Gini index for the distribution of application of applicants in the three dieses is

$$G = 1 - (3/10)^2 - (3/10)^2 - (4/10)^2 = 0.66$$

Let us now consider using each of the attributes to split the sample.

1. Attribute "Owns Home"

Value = Yes. There are five applicants who own their home. They are in classes A = 1, B = 2, C = 2.

Value = No. There are five applicants who do not own their home. They are in classes A = 2, B = l, C = 2.

Using this attribute will divide objects into those who own their home and those who do not. Computing the Gini index for each of these two subtrees,

$$G(y) = 1 - (1/5)^2 - (2/5)^2 - (2/5)^2 = 0.64$$

$$G(n) = - G(y) - 0.64$$

Total value of Gini Index $= G = 0.5G(y)\ 4 + 0.5G(n) = 0.64$

There are five applicants who are married and five that are not.

Value = Yes has A = Q = B = 1, C = 4, total 5

Value = No has A = 3, B = 2, C = 0 = total 5

Looking at the values above, it appear that this attribute will reduce the uncertainty by more than the last attribute. Computing the information gain by using this attribute, we have

$$G(y) = 1 - (1/5)^2 - (4/5)^2 = 0.32$$

$$G(n) = 1 - (3/5)^2 - (2/5)^2 = 0.48$$

Total value of Gini index $= G = 0\ 5G(y) + 0\ 5G(n) = 0.40$

3. Attribute "Gender"

There are three applicants who are male and seven are female.

Value = Male has A = 0, B = 3, C = 0, total 3

Value = Female has A = 3, B = 0, C = 4, total 7

$$G(\text{Male}) = 1 - 1 = 0$$
$$G(\text{Temale}) = 1 - (3/7)^2 - (4/7)^2 = 0.511$$

Total value of Gini Index = G = 0.3G(Male) – 0.7G(Female) = 0.358

4. Attribute "Employed"

There are eight applicants who are employed and two that are not.

Value = Yes has A = 3 B = 1, C = 4, total 8

Value = No has A = 0, B = 2, C = 0, total

$$G(y) = 1 - (3/S)^2 - (l/8)^2 - (4/8)^2 = 0.594$$
$$G(n) = 0$$

Total value of Gini Index = G = 0.8G(y) + 0.2G(n) = 0.475

5. Attribute "Credit Rating"

There are five applicants who have credit rating A and five that have B.

Value = A has A = 2, B = 1, C = 2, total 5

Value = B has A = 1, B = 2, C = 2, total 5

$$G(A) = 1 - 2(2/5)^2 - (1/5)^2 = 0.64$$
$$G(B) = G(A)$$

Total value of Gini Index = G = 0.5G(A) + 0.5G(B) = 0.64

The table below summarizes the values of the Gini Index obtained for the following five attributes:

- Owns Home
- Employed
- Married
- Credit Rating
- Gender

The attribute with the largest reduction in the Gini Index is selected as the split attribute. So the split attribute is Gender.

- Gender

Attribute	*Gini Index before split*	*Gini Index after split*	*Gain*
Owns Home	0.66	0.64	0.02
Married	0.66	0.40	0.26
Gender	0.66	Q.35S	0.302
Employed	0.66	0.475	0.185
Credit Rating	0.66	0.64	0.02

The attribute with the largest reduction in the Gini Index is selected as the split attribute. So the split attribute is Gender.

4.13 OVERFITTING AND PRUNING DECISION TREE RULES

In principle, decision tree algorithm can grow each branch of the tree just deeply enough to perfectly classify the training examples. While this is sometimes a reasonable strategy, in fact it can lead to difficulties when there is noise in the data, or when the number of training examples is too small to produce a representative sample of the true target function. In either of these cases, this simple algorithm can produce trees that *over-fit* the training examples.

Over-fitting is a significant practical difficulty for decision tree learning and many other learning methods. There are several approaches to avoiding over-fitting in decision tree learning. These can be grouped into two classes:

- approaches that stop growing the tree earlier, before it reaches the point where it perfectly classifies the training data,
- approaches that allow the tree to over-fit the data, and then post prune the tree.

Although the first of these approaches might seem more direct, the second approach of post-pruning over-fit trees has been found to be more successful in practice. This is due to the difficulty in the first approach of estimating precisely when to stop growing the tree.

Regardless of whether the correct tree size is found by stopping early or by post-pruning, a key question is what criterion is to be used to determine the correct final tree size. Approaches include:

- Use a separate set of examples, distinct from the training examples, to evaluate the utility of post-pruning nodes from the tree.
- Use all the available data for training, but apply a statistical test to estimate whether expanding (or pruning) a particular node is likely to produce an improvement beyond the training set.
- Use an explicit measure of the complexity for encoding the training examples and the decision tree, halting growth of the tree when this encoding size is minimized. This approach is based on a heuristic called the Minimum Description Length principle.

The first of the above approaches is the most common and is often referred to as a training and validation set approach. In this approach, the available data are separated into two sets of examples: a *training set*, which is used to form the learned hypothesis, and a separate *validation set*, which is used to evaluate the accuracy of this hypothesis over subsequent data and, in particular, to evaluate the impact of pruning this hypothesis.

Pruning to avoid overfitting

The approach to constructing decision trees usually involves using greedy heuristics (such as Entropy reduction) that overfit the training data and lead to poor accuracy in future predictions.

In response to the problem of overfitting nearly all modern decision tree algorithms adopt a pruning strategy of some sort. Many algorithms use a technique known as postpruning or backward pruning. This essentially involves growing the tree from a dataset until all possible leaf nodes have been reached (*i.e.*, purity) and then removing particular substrees. Studies have shown that post-pruning will result in smaller and more accurate trees by up to 25%. Different pruning techniques have been developed which have been compared in several papers and like with the different splitting criteria it has been found that there is not much variation in terms of performance. There are quite a few methods that have been developed.

Pruning strategies

At each node in a tree it is possible to see the number of instances that are misclassified on a testing set by propagating errors upwards from leaf nodes. This can be compared to the error-rate if the node was replaced by the most common class resulting from that node. If the difference is a reduction in error, then the subtree at the node can be considered for pruning. This calculation is performed for all nodes in the tree and whichever one has the highest reduced-error rate is pruned. The procedure is then recursed over the freshly pruned tree until there is no possible reduction in error rate at any node.

Although now we have a re-substitution error on the training set, the errors on the testing set are reduced. This is desirable in terms of accuracy and tree size.

Other stategies to avoid overfitting

Pruning, or post-pruning as we have described here, is not the only method that has been used to avoid overfitting. Various other post-pruning methods exist, including strategies that convert the tree to rules before pruning. Recent work has involved trying to incorporate some overfitting-prevention bias into the splitting part of the algorithm. One example of this is based on the **minimum-description length** principle that states that the best hypothesis is the one that minimises length of encoding of the hypothesis and data. This has been shown to produce accurate trees with small size.

4.14 IMPROVING ACCURACY OF CLASSIFICATION METHODS

Bootstrapping, bagging and boosting are techniques for improving the accuracy of classification results. They have been shown to be very successful for certain models, for example, decision trees. All three involve combining several classification results from the same training data that has been perturbed in some way. The aim of building several decision trees by using training data that has been perturbed is to find out how these decision trees differ from those that have been obtained earlier.

The bootstrapping method can be shown that all the samples selected are somewhat different from each other since, on the average, only 63.2% of the objects in them are unique. The bootstrap samples are then used for building decision trees which are then combined to form a single decision tree.

Bagging (the name is derived from Bootstrap and aggregating) combines classification results from multiple models or results of using the same method on several different sets of training data. Bagging may also be used to improve the stability and accuracy of a complex classification model with limited training data by using subsamples obtained by resampling, with replacement, for generating models.

Bagging essentially involves simple voting (with no weights), so the final model is the one that is predicted by the majority of the trees. The different decision trees obtained during bagging should not be so different if the training data is good and large enough. If the trees obtained are very different, it only indicates instability, perhaps due to the training data being random. In such a situation, bagging will often provide a better result than using the training data only once to build a decision tree.

There is a lot of literature available on bootstrapping, bagging, and boosting. This brief introduction only provides a glimpse into these techniques but some of the points made in the literature regarding the benefits of these methods are:

- These techniques can provide a level of accuracy that usually cannot be obtained by a large single-tree model.
- Creating a single decision tree from a collection of trees in bagging and boosting is not difficult.
- These methods can often help in avoiding the problem of overfitting since a number of trees based on random samples are used.
- Boosting appears to be on the average better than bagging although it is not always so.

On some problems bagging does better than boosting.

4.15 EVALUATION CRITERIA FOR CLASSIFICATION METHODS

The criteria for evaluation of classification methods are as follows:

1. Speed
2. Robustness
3. Scalability
4. Interpretability
5. Goodness of the model
6. Flexibility
7. Time complexity

Speed

Speed involves not just the time or computation cost of constructing a model (*e.g.* a decision tree), it also includes the time required to learn to use the model. Obviously, a user wishes to minimize both times although it has to be understood that any significant data mining project will take time to plan and prepare the data. If the problem to be solved is large, a careful study of the methods available may need to be carried out so that an efficient classification method may be chosen.

Robustness

Data errors are common, in particular when data is being collected from a number of sources and errors may remain even after data cleaning. It is therefore desirable that a method be able to produce good results in spite of some errors and missing values in datasets.

Scalability

Many data mining methods were originally designed for small datasets. Many have been modified to deal with large problems. Given that large datasets are becoming common, it is desirable that a method continues to work efficiently for large disk-resident databases as well.

Interpretability

A data mining professional is to ensure that the results of data mining are explained to the decision makers. It is therefore desirable that the end-user be able to understand and gain insight from the results produced by the classification method.

Goodness of the Model

For a model to be effective, it needs to fit the problem that is being solved. For example, in a decision tree classification, it is desirable to find a decision tree of the "right" size and compactness with high accuracy.

SUMMARY

- The objective of a classifier is not to explore the data to discover interesting segments, but rather to decide how new records should be classified.
- A common approach for classifiers is to use decision trees to partition and segment records.
- Classification is a data mining function that assigns items in a collection to target categories or classes.

- The goal of classification is to accurately predict the target class for each case in the data.
- A classification task begins with a data set in which the class assignments are known.
- Classification models are tested by comparing the predicted values to known target values in a set of test data.
- The historical data for a classification project is typically divided into two data sets: one for building the model; the other for testing the model.
- Classification has many applications in customer segmentation, business modeling, marketing, credit analysis, and biomedical and drug response modeling.
- Decision trees automatically generate rules, which are conditional statements that reveal the logic used to build the tree.
- Naive Bayes uses Bayes' Theorem, a formula that calculates a probability by counting the frequency of values and combinations of values in the historical data.
- A decision tree can be used to classify an example by starting at the root of the tree and moving through it until a leaf node, which provides the classification of the instance.
- Decision tree induction is a typical inductive approach to learn knowledge on classification.
- The decision tree method is a popular and relatively simple supervised classification method that involves each node of the tree specifying a test of some attribute and each branch from the node corresponding to one of the values of the attribute.
- Over-fitting is a significant practical difficulty for decision tree learning and many other learning methods.

REVIEW QUESTIONS

1. What kind of data is the decision tree method most suitable for?
2. Briefly outline the major steps of the algorithm to construct a decision tree.
3. What is overfitting of a decision tree? What problems can overfitting lead to?
4. List five criteria for evaluating the classification methods. Discuss them briefly.
5. Describe three methods of estimating accuracy of a classification method.
6. Explain the terms bootstrapping, bagging, and boosting for improving the accuracy of a classification method.
7. What is the difference between bootstrapping, bagging and boosting?

CHAPTER 5

Cluster Analysis

The term *cluster analysis*, first used by Tryon, 1939, encompasses a number of different algorithms and methods for grouping objects of similar kind into respective categories. A general question facing researchers in many areas of inquiry is how to organize observed data into meaningful structures, that is, to develop taxonomies. In other words cluster analysis is an exploratory data analysis tool which aims at sorting different objects into groups in a way that the degree of association between two objects is maximal if they belong to the same group and minimal otherwise. Given the above, cluster analysis can be used to discover structures in data without providing an explanation or interpretation. In other words, cluster analysis simply discovers structures in data without explaining why they exist.

We deal with clustering in almost every aspect of daily life. For example, a group of diners sharing the same table in a restaurant may be regarded as a cluster of people. In food stores items of similar nature, such as different types of meat or vegetables are displayed in the same or nearby locations. There are a countless number of examples in which clustering plays an important role. For instance, biologists have to organize the different species of animals before a meaningful description of the differences between animals is possible. According to the modern system employed in biology, man belongs to the primates, the mammals, the amniotes, the vertebrates, and the animals. In this classification, the higher the level of aggregation the less similar is the members in the respective class. Man has more in common with all other primates (*e.g.,* apes) than it does with the more "distant" members of the mammals (*e.g.,* dogs), etc. In short, whatever the nature of your business is, sooner or later you will run into a clustering problem of one form or another.

Cluster analysis divides data into meaningful or useful groups (clusters). If meaningful clusters are the goal, then the resulting clusters should capture the "natural" structure of the data. For example, cluster analysis has been used to group related documents for browsing, to find genes and proteins that have similar functionality, and to provide a grouping of spatial locations prone to earthquakes. However, in other cases, cluster analysis is only a useful starting point for other purposes, *e.g.,* data compression or efficiently finding the nearest neighbors of points. Whether for understanding or utility, cluster analysis has long been used in a wide variety of fields: psychology and other social sciences, biology, statistics, pattern recognition, information retrieval, machine learning, and data mining.

5.1 INTRODUCTION

Cluster analysis classifies a set of observations into two or more mutually exclusive *unknown* groups based on combinations of interval variables. The purpose of cluster analysis is to discover a system of organizing observations, usually people, into groups where members of the groups share properties in common. It is cognitively easier for people to predict behavior or properties of people or objects based on group membership, all of whom share similar properties. It is generally cognitively difficult to deal with individuals and predict behavior or properties based on observations of other behaviors or properties.

For example, a person might wish to predict how an animal would respond to an invitation to go for a walk. He or she could be given information about the size and weight of the animal, top speed, average number of hours spent sleeping per day, and so forth and then combine that information into a prediction of behavior. Alternatively, the person could be told that an animal is either a cat or a dog. The latter information allows a much broader range of behaviors to be predicted. The trick in cluster analysis is to collect information and combine it in ways that allow classification into useful groups, such as dog or cat.

Cluster analysis classifies unknown groups while discriminant function analysis classifies known groups. The procedure for doing a discriminant function analysis is well established. There are few options, other than type of output, that need to be specified when doing a discriminant function analysis. Cluster analysis, on the other hand, allows many choices about the nature of the algorithm for combining groups. Each choice may result in a different grouping structure.

Cluster analysis has proven to be very useful in marketing. Larson (1992) describes the efforts of a company, Clarita's, to cluster neighborhoods (zip codes) into forty different groups based on census information, such as population density, income, and age. These groups were eventually given names like "Blue Blood Estates," "Shotguns and pickups," and "Bohemian Mix." This classification scheme, call PRIZM for Potential Rating Index for Zip Markets, has proven to be very useful in direct mail advertising, radio station formats, and decisions about where to locate stores.

Steps in Doing a Cluster Analysis

A common approach to doing a cluster analysis is to first create a table of relative similarities or differences between all objects and second to use this information to combine the objects into groups. The table of relative similarities is called a proximities matrix. The method of combining objects into groups is called a clustering algorithm. The idea is to combine objects that are similar to one another into separate groups.

The Proximities Matrix

Cluster analysis starts with a data matrix, where objects (usually people in the social sciences) are rows and observations are columns. From this beginning, a table is constructed where objects are both rows and columns and the numbers in the table are measures of similarity or differences between the two observations.

For example, given the following data matrix:

	X_1	X_2	X_3	X_4	X_5
O_1 O_2 O_3 O_4					

A proximities matrix would appear as follows:

	O_1	O_2	O_3	O_4
O_1				
O_2				
O_3				
O_4				

The difference between a proximities matrix in cluster analysis and a correlation matrix is that a correlation matrix contains similarities between variables (X_1, X_2) while the proximities matrix contains similarities between observations (O_1, O_2).

The researcher has dual problems at this point. The first is a decision about what variables to collect and include in the analysis. Selection of irrelevant measures will not aid in classification. For example, including the number of legs an animal has would not help in differentiating cats and dogs, although it would be very valuable in differentiating between spiders and insects.

The second problem is how to combine multiple measures into a single number, the similarity between the two observations. This is the point where univariate and multivariate cluster analysis separate. Univariate cluster analysis groups are based on a single measure, while multivariate cluster analysis is based on multiple measures.

Univariate Measures

A simpler version of the problem of how to combine multiple measures into a measure of difference between objects is how to combine a single observation into a measure of difference between objects. Consider the following scores on a test for four students:

Student	*Score*
Julie	11
John	11
Ryan	13
Bob	18

The proximities matrix for these four students would appear as follows:

	Julie	*John*	*Ryan*	*Bob*
Julie				
John				
Ryan				
Bob				

The entries of this matrix will be described using a capital "D", for distance with a subscript describing which row and column. For example, D_{34} would describe the entry in row 3, column 4, or in this case, the intersection of Ryan and Bob.

One means of filling in the proximities matrix is to compute the absolute value of the difference between scores. For example, the distance, D_{34}, between Ryan and Bob would be $| 13–18 |$ or 5. Completing the proximities matrix using the example data would result in the following:

	Julie	*John*	*Ryan*	*Bob*
Julie	0	0	2	7
John	0	0	2	7
Ryan	2	2	0	5
Bob	7	7	5	0

A second means of completing the proximities matrix is to use the squared difference between the two measures. Using the example above, D_{34}, the distance between Ryan and Bob, would be $(13–18)^2$ or 25. This distance measure has the advantage of being consistent with many other statistical measures, such as variance and the least squares criterion, and will be used in the examples that follow. The example proximities matrix using squared differences as the distance measure is presented below.

	Julie	*John*	*Ryan*	*Bob*
Julie	0	0	4	49
John	0	0	4	49
Ryan	4	4	0	25
Bob	49	49	25	0

Note that both example proximities matrices are symmetrical. Symmetrical means that row and column entries can be interchanged or that the numbers are the same on each half of the matrix defined by a diagonal running from top left to bottom right.

Multivariate Measures

When more than one measure is obtained for each observation, then some method of combining the proximities matrices for different measures must be found. Usually the matrices are summed in a combined matrix. For example, given the following scores.

	X_1	X_2
O_1	25	11
O_2	33	11
O_3	34	13
O_4	35	18

The two proximities matrices resulting from squared Euclidean distance that result could be summed to produce a combined distance matrix.

O_1	O_2	O_3	O_4	
O_1	0	64	81	100
O_2	64	0	1	4
O_3	81	1	0	1
O4	100	4	1	0

+

	O_1	O_2	O_3	O_4
O_1	0	0	4	49
O_2	0	0	4	49
O_3	4	4	0	25
O_4	49	49	25	0

=

	O_1	O_2	O_3	O_4
O_1	0	64	85	149
O_2	64	0	5	53
O_3	85	5	0	26
O_4	149	53	26	0

Note that each corresponding cell is added. With more measures there are more matrices to be added together.

This system works reasonably well if the measures share similar scales. One measure can overwhelm the other if the measures use different scales. Consider the following scores.

	X_1	X_2
O_1	25	11
O_2	33	21
O_3	34	33
O_4	35	48

The two proximities matrices resulting from squared Euclidean distance that result could be summed to produce a combined distance matrix.

	O_1	O_2	O_3	O_4
O_1	0	64	81	100
O_2	64	0	1	4
O_3	81	1	0	1
O_4	100	4	1	0

+

	O_1	O_2	O_3	O_4
O_1	0	100	484	49
O_2	100	0	144	729
O_3	484	144	0	225
O_4	1369	729	225	0

=

	O_1	O_2	O_3	O_4
O_1	0	164	485	153
O_2	164	0	145	733
O_3	565	145	0	226
O_4	1469	733	226	0

It can be seen that the second measure overwhelms the first in the combined matrix.

For this reason the measures are optionally transformed before they are combined. For example, the previous data matrix might be converted to standard scores before computing the separated distance matrices.

	X_1	X_2	Z_1	Z_2
O_1	25	11	–1.48	–1.08
O_2	33	21	.27	–.45
O_3	34	33	.49	.30
O_4	35	48	.71	1.24

The two proximities matrices resulting from squared Euclidean distance that result from the standard scores could be summed to produce a combined distance matrix.

	O_1	O_2	O_3	O_4
O_1	0	3.06	3.88	4.80
O_2	3.06	0	.05	.19
O_3	3.88	.05	0	.05
O_4	4.80	.19	.05	0

+

	O_1	O_2	O_3	O_4
O_1	0	.40	1.90	5.38
O_2	.40	0	.56	2.86
O_3	1.9	.56	0	.88
O_4	5.38	2.86	.88	0

=

	O_1	O_2	O_3	O_4
O_1	0	3.46	5.78	10.18
O_2	3.46	0	.61	3.05
O_3	5.78	.61	0	.93
O_4	10.18	3.05	.93	0

The point is that the choice of whether to transform the data and the choice of distance metric can result in vastly different proximities matrices.

Using Distances to Group Objects

After the distances between objects have been found, the next step in the cluster analysis procedure is to divide the objects into groups based on the distances. Again, any number of options are available to do this.

If the number of groups is known beforehand, a "flat" method might be preferable. Using this method, the objects are assigned to a given group at the first step based on some initial criterion. The means for each group are calculated. The next step reshuffles the objects into groups, assigning objects to groups based on the object's similarity to the current mean of that group. The means of the groups are recalculated at the end of this step. This process continues recursively until no objects change groups. This idea is the basis for "k-means cluster analysis" available on SPSS/WIN and other statistical packages. This method works well if the number of groups match the data and the initial solution is reasonably close to the final solution.

Hierarchical clustering methods do not require preset knowledge of the number of groups. Two general methods of hierarchical clustering methods are available: divisive and agglormerative.

The divisive techniques start by assuming a single group, partitioning that group into subgroups, partitioning these subgroups further into subgroups and so on until each object forms its own subgroup. The agglormerative techniques start with each object describing a subgroup, and then combine like subgroups into more inclusive subgroups until only one group remains. The results of the application of the clustering technique are best described using a dendogram or binary tree. The objects are represented as nodes in the dendogram and the branches illustrate when the cluster method joins subgroups containing that object. The length of the branch indicates the distance between the subgroups when they are joined.

Single Linkage

Single linkage (nearest neighbor in SPSS/WIN) computes the distance between two subgroups as the minimum distance between any two members of opposite groups. For example, consider the following proximities matrix.

	Julie	*John*	*Ryan*	*Bob*	*Ted*	*Kristi*
Julie	0	4	36	81	196	225
John	4	0	16	49	144	169
Ryan	36	16	0	9	64	81
Bob	81	49	9	0	25	36
Ted	196	144	64	25	0	1
Kristi	225	169	81	36	1	0

Based on this data, Ted and Kristi would be joined on the first step.

	Julie	*John*	*Ryan*	*Bob*	*Ted and Kristi*
Julie	0	4	36	81	196
John	4	0	16	49	144
Ryan	36	16	0	9	64
Bob	81	49	9	0	25
Ted and Kristi	196	144	64	25	0

Julie and John would be joined on the second step.

	Julie and John	*Ryan*	*Bob*	*Ted and Kristi*
Julie and John	0	16	49	144
Ryan	16	0	9	64
Bob	49	9	0	25
Ted and Kristi	144	64	25	0

Ryan and Bob would be joined on the third step. At that step four would be three groups of two each and the proximities matrix at that point would appear as follows.

	Julie and John	*Ryan and Bob*	*Ted and Kristi*
Julie and John	0	16	144
Ryan and Bob	16	0	25
Ted and Kristi	144	25	0

Using single linkage, Julie and John would be grouped with Ryan and Bob at the third step. The distance of this linkage would be 16. The last step would join all into a single group with a distance of 25.

Complete Linkage

Complete linkage (furthest neighbor in SPSS/WIN) computes the distance between subgroups in each step as the maximum distance between any two members of the different groups. Using the example proximities matrix, complete linkage would join similar members at steps 1, 2, and 3 in the procedure. At step four the proximities matrix for the three groups would appear as follows.

	Julie and John	*Ryan and Bob*	*Ted and Kristi*
Julie and John	0	81	225
Ryan and Bob	81	0	81
Ted and Kristi	225	81	0

Step four would combine all three subgroups into a single group with a distance of 81.

Average Linkage

Average linkage (or Centroid Method in SPSS/WIN) computes the distance between subgroups at each step as the average of the distances between the two subgroups.

Using the example proximities matrix, average linkage would join similar members at steps 1, 2, and 3 in the procedure. At step four the proximities matrix for the three groups would appear as follows.

	Julie and John	*Ryan and Bob*	*Ted and Kristi*
Julie and John	0	45.5	51.5
Ryan and Bob	45.5	0	81
Ted and Kristi	183.5	51.5	0

Thus, Julie and John would be grouped with Ryan and Bob with a distance of 51.5. The final step would join this group with Ted and Kristi with an average distance of 117.5.

A program generating a cluster analysis homework assignment permits the student to view a dendogram for single, complete, and average linkage for a random proximities matrix. The student should verify that different dendograms result when different linkage methods are used.

Cluster analysis methods will always produce a grouping. The groupings produced by cluster analysis may or may not prove useful for classifying objects. If the groupings discriminate between variables not used to do the grouping and those discriminations are useful, then cluster analysis is useful. For example, if grouping zip code areas into fifteen categories based on age, gender, education, and income discriminates between wine drinking behaviors, it would be very useful information if one was interested in expanding a wine store into new areas.

Cluster analysis may be used in conjunction with discriminant function analysis. After multivariate data are collected, observations are grouped using cluster analysis. Discriminant function analysis is then used on the resulting groups to discover the linear structure of either the measures used in the cluster analysis and/or different measures.

Cluster analysis methods are not clearly established. There are many options one may select when doing a cluster analysis using a statistical package. Cluster analysis is thus open to the criticism that a statistician may mine the data trying different methods of computing the proximities matrix and linking groups until he or she "discovers" the structure that he or she originally believed was contained in the data. One wonders why anyone would bother to do a cluster analysis for such a purpose.

5.2 TYPES OF DATA

Here we will study about different types of data that often occurs in cluster analysis.

(*a*) **Interval Scaled Variables:** These are the continuous measurements of a roughly linear scale. The measurement unit used can affect clustering analysis. For example, converting meters to inches or weights to pounds can result in a very different clustering structure. Expressing a variable in smaller units will lead to a larger range for that variable in smaller units will lead to a larger range for that variable, and thus a larger effect on the resulting clustering structure.

(*b*) **Binary Variables:** a binary variable has only two states: 0 and 1, where 0 marks the absence of a variable and 1 marks the presence of a variable. Given the Boolean value for a person as ill, 0 marks the person as not ill while 1 marks the person as ill. Boolean variables are of two types: symmetric and asymmetric. A variable is symmetric if both of its states are equally variable and carry the same weight that is, there is no preference on which outcome should be coded as 0 or 1. A binary variable is asymmetric if the outcome of the states is not equally important such as positive or negative outcome of a disease.

(*c*) **Categorical Variables:** It is a generation of binary variables in that it can take on more than two states. For *e.g.*, the states in a map may be marked in different colors.

(*d*) **Ordinal Variables:** These are very useful for registering subjective assessments of qualities that cannot be measured objectively.

(*e*) **Ratio scaled Variables:** These make a positive measurement on a nonlinear scale such as exponents.

5.3 DESIRED FEATURES OF CLUSTER ANALYSIS

The desired characteristics of a clustering algorithm depend on the particular problem under consideration. The following is a list.

- **Scalability:** Clustering techniques for large sets of data must be scalable, both in terms of speed and space. It is not unusual for a database to contain millions of records, and thus, any clustering algorithm used should have linear or near linear time complexity to handle such large data sets. (Even algorithms that have complexity of $O(m^2)$ are not practical for large data sets.) Some clustering techniques use statistical sampling. Nonetheless, there are cases, *e.g.*, situations where relatively rare points have a dramatic effect on the final clustering, where a sampling is insufficient. Furthermore, clustering techniques for databases cannot assume that all the data will fit in main memory or that data elements can be randomly accessed. These algorithms are, likewise, infeasible for large data sets. Accessing data points sequentially and not being dependent on having all the data in main memory at once are important characteristics for scalability.

- Independence of the order of input: Some clustering algorithms are dependent on the order of the input, *i.e.*, if the order in which the data points are processed changes, then the resulting clusters may change. This is unappealing since it calls into question the validity of the clusters that have been discovered. They may just represent local minimums or artifacts of the algorithm.
- **Effective means of detecting and dealing with noise or outlying points:** A point which is noise or is simply an atypical point (outlier) can often distort a clustering algorithm. By applying tests that determine if a particular point really belongs to a given cluster, some algorithms can detect noise and outliers and delete them or otherwise eliminate their negative effects. This processing can occur either while the clustering process is taking place or as a post-processing step. However, in some instances, points cannot be discarded and must be clustered as well as possible. In such cases, it is important to make sure that these points do not distort the clustering process for the majority of the points.
- **Effective means of evaluating the validity of clusters that are produced:** It is common for clustering algorithms to produce clusters that are not "good" clusters when evaluated later.
- **Easy interpretability of results:** Many clustering methods produce cluster descriptions that are just lists of the points belonging to each cluster. Such results are often hard to interpret. A description of a cluster as a region may be much more understandable than a list of points. This may take the form of a hyper-rectangle or a center point with a radius. Also, data clustering is sometimes preceded by a transformation of the original data space - often into a space with a reduced number of dimensions. While this can be helpful for finding clusters, it can make the results very hard to interpret.
- **The ability to find clusters in subspaces of the original space:** Clusters often occupy a subspace of the full data space. (Hence, the popularity of dimensionality reduction techniques.) Many algorithms have difficulty finding, for example, a 5 dimensional cluster in a 10 dimensional space.
- The ability to handle distances in high dimensional spaces properly: High-dimensional spaces are quite different from low dimensional spaces. It has been proved that the distances between the closest and farthest neighbors of a point may be very similar in high dimensional spaces. Perhaps an intuitive way to see this is to realize that the volume of a hyper-sphere with radius, *r*, and dimension, *d*, is proportional to *rd* , and thus, for high dimensions a small change in radius, means a large change in volume. Distance based clustering approaches may not work well in such cases. If the distance between points in a high dimensional space is plotted, then the graph will often show two peaks: a "small" distance representing the distance between points in clusters, and a "larger" distance representing the average distance between points. If only one peak is present or if the two peaks are close, then clustering via distance based approaches will likely be difficult. Yet another set of problems has to do with how to weight the different dimensions. If different aspects of the data are being measured in different scales, then a number of difficult issues arise. Most distance functions will weight dimensions with greater ranges of data more highly. Also, clusters that are determined by

using only certain dimensions may be quite different from the clusters determined by using different dimensions. Some techniques are based on using the dimensions that result in the greatest differentiation between data points. Many of these issues are related to the topic of feature selection, which is an important part of pattern recognition.

- **Robustness in the presence of different underlying data and cluster characteristics:** A robust clustering technique must be take into account the following characteristics of the data and the clusters:

  (*a*) Dimensionality

  (*b*) Noise and Outliers

  (*c*) Statistical Distribution

  (*d*) Cluster Shape

  (*e*) Cluster Size

  (*f*) Cluster Density

  (*g*) Cluster Separatio

  (*h*) Type of data space, *e.g.,* Euclidean or non-Euclidean

  (*i*) Many and Mixed Attribute Types, *e.g.*, (temporal, continuous, categorical).

A mix of attribute types is usually handled by having a proximity function that can combine all the different attributes in an "intelligent" way. Many clustering techniques have assumption about the data and cluster characteristics and do not work well if those assumptions are violated. In such cases the clustering technique may miss clusters, split clusters, merge clusters, or just produce poor clusters.

- **An ability to estimate any parameters:** Many clustering algorithms take the number of clusters as a parameter. This can be a useful feature in certain instances, *e.g.,* when using a clustering algorithm to create a balanced tree for nearest neighbor lookup, or when using clustering for compression. However, this is not generally good, since the number of clusters parameterized may not match the "real" number of clusters. An algorithm that requires the number of clusters up front can always be run multiple times. Assuming that there is some way to compare the quality of the clusters produced, it is then possible to empirically determine the best number of clusters. Of course, this increases the amount of computation required. Likewise, it is often difficult to estimate the proper values for other parameters of clustering algorithms. In general, the parameters of a clustering algorithm may identify areas of weakness. In the best case, the results produced by a clustering algorithm will be relatively insensitive to modest changes in the parameter values.
- **Ability to function in an incremental manner:** In certain cases, *e.g.*, data warehouses, the underlying data used for the original clustering can change over time. If the clustering algorithm can incrementally handle the addition of new data or the deletion of old data, then this is usually much more efficient than re-running the algorithm on the new data set.

5.4 TYPES OF CLUSTER ANALYSIS METHODS

Many different clustering techniques have been proposed over the years. These techniques can be described using the following criteria:

A. Hierarchical vs. partitional (nested and unnested). Hierarchical techniques produce a nested sequence of partitions, with a single, all inclusive cluster at the top and singleton clusters of individual points at the bottom. Each intermediate level can be viewed as combining (splitting) two clusters from the next lower (next higher) level. (While most hierarchical algorithms involve joining two clusters or splitting a cluster into two sub-clusters, some hierarchical algorithms join more than two clusters in one step or split a cluster into more than two sub-clusters.) The following figures indicate different ways of graphically viewing the hierarchical clustering process. Figures 5.1 and 5.2 illustrate the more "traditional" view of hierarchical clustering as a process of merging two clusters or splitting one cluster into two. Figure 5.1 gives a "nested set" representation of the process, and a "tree' representation, or dendogram. Figure 5.2 show a different, hierarchical clustering, one in which points $p1$ and $p2$ are grouped together in the same step that also groups points $p3$ and $p4$ together.

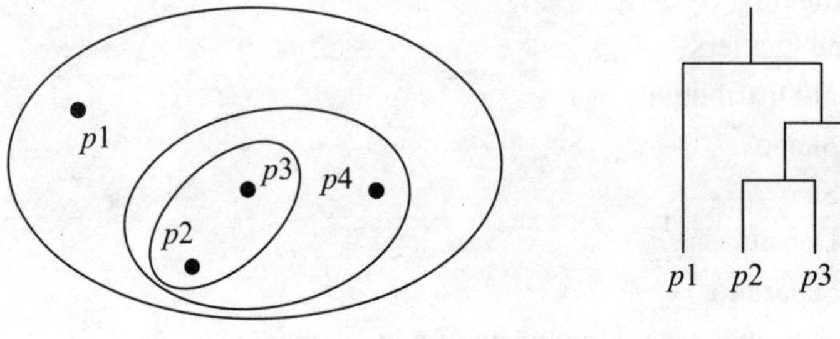

Figure 5.1. Traditional Nested set and Dendogram

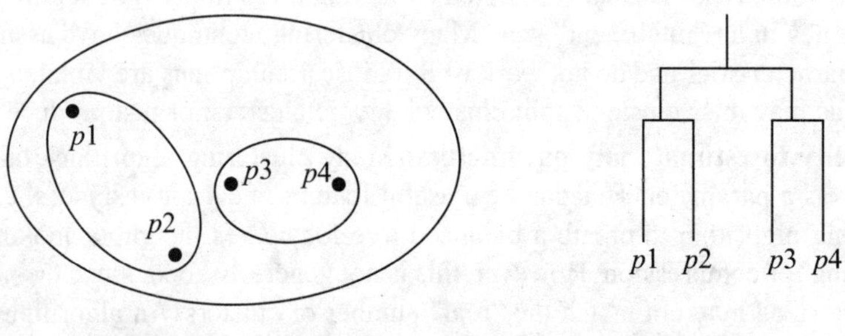

Figure 5.2. Non-traditional Nested set and Dendogram

Partitional techniques create a one-level (unnested) partitioning of the data points. If K is the desired number of clusters, then partitional approaches typically find all K clusters at once. Contrast this with traditional hierarchical schemes, which bisect a cluster to get two clusters or merge two clusters to get one. Of course, a hierarchical approach can be used to generate a flat partition of K clusters, and likewise, the repeated application of a partitional scheme can provide a hierarchical clustering.

B. Divisive vs. agglomerative. Hierarchical clustering techniques proceed either from the top to the bottom or from the bottom to the top, *i.e.*, a technique starts with one large cluster and splits it, or starts with clusters each containing a point, and then merges them.

C. Incremental or non-incremental. Some clustering techniques work with an item at a time and decide how to cluster it given the current set of points that have already been processed. Other techniques use information about all the points at once. Non-incremental clustering algorithms are far more common.

5.5 PARTITIONAL METHODS

Partitional clustering techniques create a one-level partitioning of the data points. There are a number of such techniques, but we shall only describe two approaches in this section: K-means and K-medoid. Both these techniques are based on the idea that a center point can represent a cluster. For K-means we use the notion of a centroid, which is the mean or median point of a group of points. Note that a centroid almost never corresponds to an actual data point. For K-medoid we use the notion of a medoid, which is the most representative (central) point of a group of points. By its definition a medoid is required to be an actual data point.

Wealso discuss CLARANS, a more efficient version of the basic K-medoid that was applied to spatial data mining problems. BIRCH and Bubble are clustering methods for data mining that are based on center-based partitional approaches. (BIRCH is more like K-means and Bubble is more like K-medoid.) However, both of BIRCH and Bubble use a wide variety of techniques to help enhance their scalability, *i.e.*, their ability to handle large data sets.

K-means Clustering

The K-means clustering technique is very simple and we immediately begin with a description of the basic algorithm. We elaborate in the following sections. Basic K-means Algorithm for finding K clusters.

1. Select K points as the initial centroids.
2. Assign all points to the closest centroid.
3. Recompute the centroid of each cluster.
4. Repeat steps 2 and 3 until the centroids don't change.

In the absence of numerical problems, this procedure always converges to a solution, although the solution is typically a local minimum. The following diagram gives an example of this. Figure 3*a* shows the case when the cluster centers coincide with the circle centers. This is a global minimum. Figure 5.3(*b*) shows a local minima.

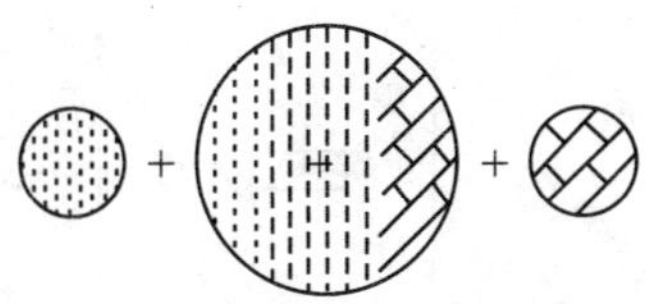

Figure 5.3. (*a*) A globally minimal clustering solution

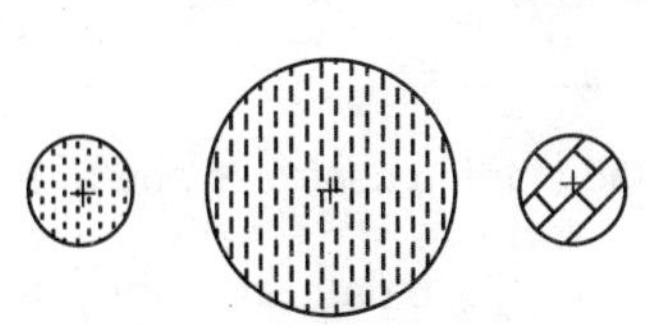

Figure 5.3. (*b*) A locally minimal clustering solution.

Time and Space Complexity

Since only the vectors are stored, the space requirements are basically O(mn), where m is the number of points and *n* is the number of attributes. The time requirements are O(I*K**m***n*), where I is the number of iterations required for convergence. I is typically small (5-10) and can be easily bounded as most changes occur in the first few iterations. Thus, K-means is linear in *m*, the number of points, and is efficient, as well as simple, as long as the number of clusters is significantly less than *m*.

Choosing Initial Centroids

Choosing the proper initial centroids is the key step of the basic K-means procedure. It is easy and efficient to choose initial centroids randomly, but the results are often poor. It is possible to perform multiple runs, each with a different set of randomly chosen initial centroids - one study advocates 30 - but this may still not work depending on the data set and the number of clusters sought. We start with a very simple example of three clusters and 16 points. Figure 5.4(*a*) indicates the "natural" clustering that results when the initial centroids are "well" distributed. Figure 5.4(*b*) indicates a "less natural" clustering that happens when the initial centroids are poorly chosen.

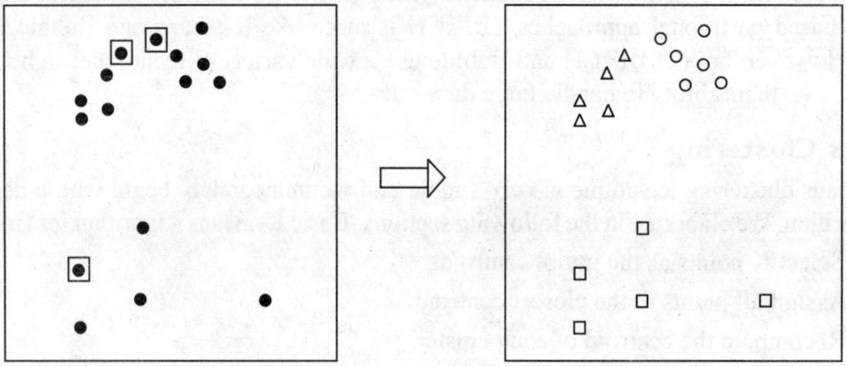

Figure 5.4. (*a*) Good starting centroids and a "natural" clustering.

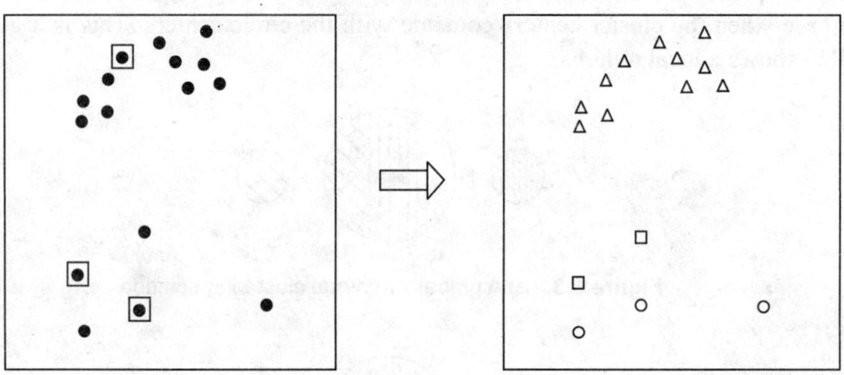

Figure 5.4. (*b*) Bad starting centroids and a "less natural" clustering.

We have also constructed the artificial data set, shown in Figure 5.5(*a*) as another illustration of what can go wrong. The figure consists of 10 pairs of circular clusters, where each cluster of a pair of clusters is close to each other, but relatively far from the other clusters. The probability that an initial centroid will come from any given cluster is 0.10, but the probability that each cluster will have exactly one initial centroid is 10!/101018 = 0.00036.

There isn't any problem as long as two initial centroids fall anywhere in a pair of clusters, since the centroids will redistribute themselves, one to each cluster, and so achieve a globally minimal error,. However, it is very probable that one pair of clusters will have only one initial centroid. In that case, because the pairs of clusters are far apart, the K-means algorithm will not redistribute the centroids between pairs of clusters, and thus, only a local minima will be achieved. When starting with an uneven distribution of initial centroids as shown in figure 5.5(*b*), we get a non-optimal clustering, as is shown in figure 5.5(*c*), where different fill patterns indicate different clusters. One of the clusters is split into two clusters, while two clusters are joined in a single cluster.

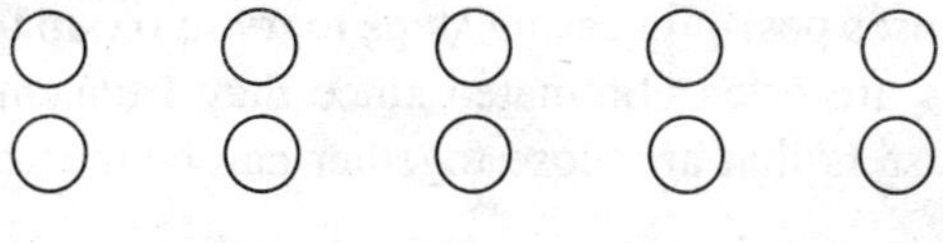

Figure 5.5. (*a*) Data distributed in 10 circular regions

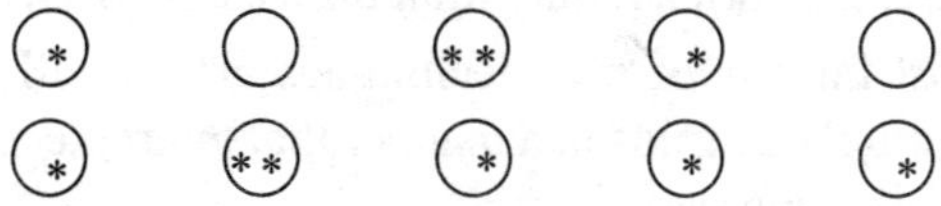

Figure 5.5. (*b*) Initial Centroids

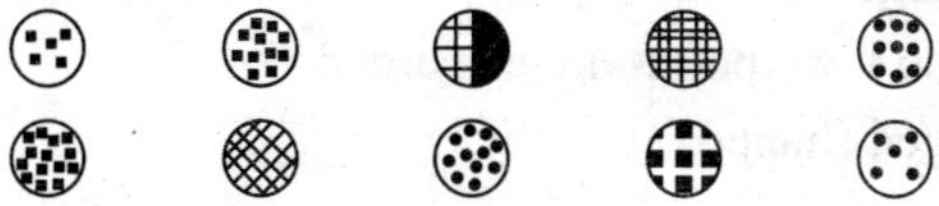

Figure 5.5. (*c*) K-means clustering result

Because random sampling may not cover all clusters, other techniques are often used for finding the initial centroids. For example, initial centroids are often chosen from dense regions, and so that they are well separated, *i.e.*, so that no two centroids are chosen from the same cluster.

Updating Centroids Incrementally

Instead of updating the centroid of a cluster after all points have been assigned to clusters, the centroids can be updated as each point is assigned to a cluster. In addition, the relative weight of the point being added may be adjusted. The goal of these modifications is to achieve better accuracy and faster convergence. However, it may be difficult to make a good choice for the relative weight. These update issues are similar to those of updating weights for artificial neural nets. Incremental update also has another advantage—empty clusters are not produced. All clusters start with a single point and if a cluster ever gets down to one point, then that point will always be reassigned to that cluster. Empty clusters are often observed when centroid updates are performed only after all points have been assigned to clusters.

This imposes the need for a technique to choose a new centroid for an empty cluster, for otherwise the squared error will certainly be larger than it would need to be. A common approach is to choose as the new centroid the point that is farthest away from any current center. If nothing else, this eliminates the point that currently contributes the most to the squared error. Updating points incrementally may introduce an order dependency problem, which can be ameliorated by randomizing the order in which the points are processed. However, this is not really feasible unless the points are in main memory. Updating the centroids after all points are assigned to clusters results in order independence approach.

Finally, note that when centers are updated incrementally, each step of the process may require updating two centroids if a point switches clusters. However, K-means tends to converge rather quickly and so the number of points switching clusters will tend to be small after a few passes over all the points.

Pre- and Post-Processing

Sometimes the quality of the clusters that are found can be improved by preprocessing the data. For example, when we use the squared error criteria, outliers can unduly influence the clusters that are found. Thus, it is common to try to find such values and eliminate them with a preprocessing step.

Another common approach uses post-processing steps to try to fix up the clusters that have been found. For example, small clusters are often eliminated since they frequently represent groups of outliers. Alternatively, two small clusters that are close together can be merged. Finally, large clusters can be split into smaller clusters.

While pre-and post-processing techniques can be effective, they are based on heuristics that may not always work. Furthermore, they often require that the user choose values for a number of parameters.

ISODATA is the most famous of such techniques and is still used today, especially in image processing. After choosing initial centroids in a manner that guarantees that they will be well separated, ISODATA proceeds in the following way:

(1) Points are assigned to the closest centroid and cluster centroids are recomputed. This step is repeated until no points change clusters.

(2) Clusters with "too few" points are discarded.

(3) Clusters are merged (lumped) or split.

(*a*) If there are "too many" clusters, then clusters are split.

(*b*) If there are "too few" clusters compared to the number desired, then clusters are merged.

(*c*) Otherwise the cluster splitting and merging phases alternate.

Clusters are merged if their centers are close, while clusters are split if the are "too loose."

(4) Steps 1, 2 and 3 are repeated until the results are "acceptable" or until some predetermined number of iterations has been exceeded.

The splitting phase of ISODATA is an example of strategies that decrease the total error by increasing the number of clusters. Two general strategies are given below:

(*a*) Split a cluster. Often the cluster with the largest error is chosen, but in the case of ISODATA the cluster with the largest standard deviation in some particular attribute is chosen.

(*b*) Introduce a new cluster center. Often the point that is farthest from any cluster center is chosen.

Likewise, the merging phase of ISODATA is just one example of strategies that decrease the number of clusters. (This will increase the error.) Two general strategies are given below:

(*a*) *Disperse a cluster, i.e.,* remove the center that corresponds to the center and reassign the points to other centers. Ideally the cluster which is dispersed should be the one that increases the error the least. A good heuristic for determining this is based on keeping track of the second closest center of each point.

(*b*) *Merge two clusters.* In ISODATA, the clusters with the closest centers are chosen, although another, perhaps better, approach is to merge the two clusters that result in the smallest increase in error. These two merging strategies are, respectively, the same strategies that are used in the hierarchical clustering techniques known as the Centroid method and Ward's method.

Finally, it is important to realize that there are certain clustering applications, where outliers cannot be eliminated. In data compression every point must be clustered, and in financial analysis, apparent outliers are often the most interesting points, *e.g.*, unusually profitable customers.

Limitations and Problems

K-means attempts to minimize the squared or absolute error of points with respect to their cluster centroids. While this is sometimes a reasonable criterion and leads to a simple algorithm, K-means has a number of limitations and problems. In particular, Figures 5.6(*a*) and 5.6(*b*) show the problems that result when clusters have widely different sizes or have convex shapes.

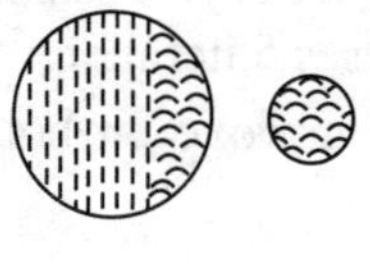

Figure 5.6. (a) Different sizes

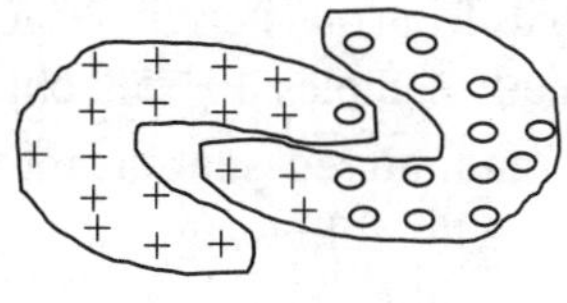

Figure 5.6. (b) Convex shapes

The difficulty in these two situations is that the K-means objective function is a mismatch for the kind of clusters we are trying to find. The K-means objective function is minimized by globular clusters of equal size or by clusters that are well separated.

The K-means algorithm is also normally restricted to data in Euclidean spaces because in many cases the required means and medians do not make sense. (This is not true in all cases, *e.g.*, documents.) A related technique, K-medoid clustering, does not have this restriction and is discussed in the next section.

K-medoid Clustering

The objective of K-medoid clustering is to find a non-overlapping set of clusters such that each cluster has a most representative point, *i.e.*, a point that is most centrally located with respect to some measure, *e.g.*, distance. These representative points are called medoids. Once again, the algorithm is conceptually simple. Basic K-medoid Algorithm for finding K clusters.

(1) Select K initial points. These points are the candidate medoids and are intended to be the most central points of their clusters.

(2) Consider the effect of replacing one of the selected objects (medioids) with one of the non-selected objects. Conceptually, this is done in the following way. The distance of each non-selected point from the closest candidate medoid is calculated, and this distance is summed over all points. This distance represents the "cost" of the current configuration. All possible swaps of a non-selected point for a selected one are considered, and the cost of each configuration is calculated.

(3) Select the configuration with the lowest cost. If this is a new configuration, then repeat step 2.

(4) Otherwise, associate each non-selected point with its closest selected point (medoid) and stop.

Step 3 requires more explanation. The ith medoid is evaluated by using $\sum_{j=1}^{n_1} p_{ij}$ where p_{ij}, is the proximity between the ith medoid and the jthpoint in the cluster. For similarities (dissimilarities) we want this sum to be as large (small) as possible. It can be seen that such an approach is not restricted to Euclidean spaces and is likely to be more tolerant of outliers. However, finding a better medoid requires trying all points that are currently not medoids and is computationally expensive.

CLARANS

CLARANS was one of the first clustering algorithms that was developed specifically for use in data mining - spatial data mining. CLARANS itself grew out of two clustering algorithms, PAM and CLARA, that were developed in the field of statistics. CLARA (Clustering LARge Applications) is an adaptation of PAM for handling larger data sets. It works by repeatedly sampling a set of data points, calculating the medoids of the sample, and evaluating the cost of the configuration that consists of these "sample-derived" medoids and the entire data set. The set of medoids that minimizes the cost is selected. As an optimization, the set of medoids from one iteration is usually included in the sample for the next iteration. As presented, CLARA normally samples 40+ 5K objects and uses 5 iterations.

CLARANS uses a randomized search approach to improve on both CLARA and PAM. Conceptually, CLARANS does the following.

(1) Randomly pick K candidate medoids.

(2) Randomly consider a swap of one of the selected points for a non-selected point.

(3) If the new configuration is better, i.e., has lower cost, then repeat step 2 with the new configuration.

(4) Otherwise, repeat step 2 with the current configuration unless a parameterized limit has been exceeded. (This limit was set to max (250, K *(m - K)).

(5) Compare the current solution with any previous solutions and keep track of the best.

(6) Return to step 1 unless a parameterized limit has been exceeded. (This limit was set to 2.)

Thus, CLARANS differs from PAM in that, given the current configuration, it does not consider all possible swaps of medoid and non-medoid points, but rather, only a random selection of them and only until it finds a better configuration. Unlike CLARA, CLARANS works with all data points. CLARANS was used in two spatial data mining tools, SD(CLARANS) and NSD(CLARANS). These tools added some capabilities related to cluster analysis.

5.6 HIERARCHICAL METHODS

The clustering algorithms are broadly classified into two namely hierarchical and non-hierarchical algorithms. In the hierarchical procedures, we construct a hierarchy or tree-like structure to see the relationship among entities (observations or individuals). In the non-hierarchical method a position in the measurement is taken as central place and distance is measured from such central point (seed). Identifying a right central position is a big challenge and hence non-hierarchical methods are less popular.

There is a concept of ordering involved in this approach. The ordering is driven by how many observations could be combined at a time or what determines that the distance is not statistically different from 0 between two observations or two clusters. The clusters could be arrived at either from weeding out dissimilar observations (divisive method) or joining together similar observations (agglomerative method). Most common statistical packages use agglomerative method and the most popular agglomerative methods are

- Single linkage (nearest neighbor approach),
- Complete linkage (furthest neighbor),
- Average linkage,
- Ward's method,
- Centroid method.

All these differ in the definition of distance and what defines largest distance as statistically no-distance or zero-distance. Most of the time, the distance is based on Euclidean distance in the sample axes (Mahalanobis distance is for non-orthogonal sample).

In hierarchical clustering the goal is to produce a hierarchical series of nested clusters, ranging from clusters of individual points at the bottom to an all-inclusive cluster at the top. A diagram called a dendogram graphically represents this hierarchy and is an inverted tree that describes the order in which points are merged (bottom-up view) or clusters are split (top-down view).

One of the attractions of hierarchical techniques is that they correspond to taxonomies that are very common in the biological sciences, *e.g.*, kingdom, phylum, genus, species, etc. (Some cluster analysis work occurs under the name of "mathematical taxonomy.") Another attractive feature is that hierarchical techniques do not assume any particular number of clusters. Instead any desired number of clusters can be obtained by "cutting" the dendogram at the proper level. Finally, hierarchical techniques are thought to produce better quality clusters.

Agglomeration and Division

There are two basic approaches to generating a hierarchical clustering:

(*a*) **Agglomerative:** Start with the points as individual clusters and, at each step, merge the closest pair of clusters. This requires defining the notion of cluster proximity.

(*b*) **Divisive:** Start with one, all-inclusive cluster and, at each step, split a cluster until only singleton clusters of individual points remain. In this case, we need to decide which cluster to split at each step.

Divisive techniques are less common, and we will mention only one example before focusing on agglomerative techniques.

Simple Divisive Algorithm (Minimum Spanning Tree (MST))

(1) Compute a minimum spanning tree for the proximity graph.

(2) Create a new cluster by breaking the link corresponding to the largest distance (smallest similarity).

(3) 3 Repeat step 2 until only singleton clusters remain.

This approach is the divisive version of the "single link" agglomerative technique

Basic Agglomerative Hierarchical Clustering Algorithm

Many hierarchical agglomerative techniques can be expressed by the following algorithm, which is known as the Lance-Williams algorithm. Basic Agglomerative Hierarchical Clustering Algorithm

(1) Compute the proximity graph, if necessary. (Sometimes the proximity graph is all that is available.)

(2) Merge the closest (most similar) two clusters.

(3) Update the proximity matrix to reflect the proximity between the new cluster and the original clusters.

(4) Repeat steps 3 and 4 until only a single cluster remains.

The key step of the previous algorithm is the calculation of the proximity between two clusters, and this is where the various agglomerative hierarchical techniques differ.

Any of the cluster proximities that we discuss in this section can be viewed as a choice of different parameters (in the Lance-Williams formula) for the proximity between clusters Q and R, where R is formed by merging clusters A and B.

$$p(R, Q) = \alpha A\, p(A, Q) + \alpha B\, p(B, Q) + \beta\, p(A, Q) + \gamma\, |\, p(A, Q) - p(B, Q)\, |$$

In words, this formula says that after you merge clusters A and B to form cluster R, then the distance of the new cluster, R, to an existing cluster, Q, is a linear function of the distances of Q from the original clusters A and B.

Any hierarchical technique that can be phrased in this way does not need the original points, only the proximity matrix, which is updated as clustering occurs. However, while a general formula is nice, it is often easier to understand the different hierarchical methods by looking directly at the definition of cluster distance that each method uses, and that is the approach that we shall take here.

Time and Space Complexity

Hierarchical clustering techniques typically use a proximity matrix. This requires the computation and storage of m^2 proximities, a factor that limits the size of data sets that can be processed. It is possible to compute the proximities on the fly and save space, but this increases computation time. Once the proximity matrix is available, the time required for hierarchical clustering is $O(m^2)$.

Key Issues in Hierarchical Clustering

- **Lack of a Global Objective Function:** We have already mentioned that hierarchical clustering cannot be viewed as globally optimizing an objective function. Instead, hierarchical clustering techniques use various criteria to decide "locally" at each step which clusters should be joined (or split for divisive approaches). For agglomerative hierarchical techniques, the criteria is typically to merge the "closest" pair of clusters, where "close" is defined by a specified measure of cluster proximity. This approach yields clustering algorithms that avoid the difficulty of trying to solve a hard combinatorial optimization problem. (As previously, the general clustering problem for objective functions like squared error is NP complete.) Furthermore, such approaches do not have problems with local minima or difficulties in choosing initial points. Of course, the time and space complexity is $O(m^2)$, which is prohibitive in many cases.
- **Cluster Proximity and Cluster Representation:** Cluster proximity is typically defined by a conceptual view of the clusters. For example, if we view a cluster as being represented by all the points, then we can take the proximity between the closest (most similar) two points in different clusters as being the proximity between the two clusters. This defines the MIN technique. Likewise, we can also take the proximity between the farthest (least similar) two points in different clusters as our definition of cluster proximity. This defines the MAX technique. Finally, we can average the pairwise proximities of all pairs of two points from different clusters. This yields the Group Average technique. If instead, we chose to represent each cluster by a centroid, then we find that different definitions of cluster proximity are more natural. In one case, the cluster distance is defined as the distance between cluster centroids. This defines the Centroid clustering technique. Ward's method is conceptually related to Centroid clustering since it merges the two clusters that result in the smallest increase in the total squared error, *i.e.,* it minimizes the sum of the squared distance of points from their centroids. However, Ward's method can also be viewed as an example of cluster methods that take an "all points" view of cluster representation, but define proximity in terms of preserving desired cluster properties. For Ward's method, the desired cluster property is a low squared error, *i.e.*, minimum squared distance of all points from the cluster center. Chameleon uses the concepts of cluster closeness and connectivity to define cluster properties that must be preserved.
- **The Impact of Cluster Size:** Another aspect of hierarchical agglomerative clustering that should be considered is how to treat the relative sizes of the pairs of clusters that may be merged. There are basically two schemes: weighted and unweighted. Weighted schemes treat each cluster equally and thus, objects in smaller clusters effectively have larger weight. Unweighted schemes treat all objects in the clusters equally. Note that this discussion only applies to cluster proximity schemes that involve sums, *i.e.,* Centroid and Group Average. Unweighted schemes seem to be more popular, and in our further discussions about the Group Average and Centroid schemes, we refer only to the unweighted versions.

- **Merging Decisions are Final:** Hierarchical clustering tends to make good local decisions about combining two clusters since it has the entire proximity matrix available. However, once a decision is made to merge two clusters, the hierarchical scheme does not allow for that decision to be changed. This prevents a local optimization criterion from becoming a global optimization criterion. For example, in Ward's method, the "minimize squared error" criteria from Kmeans is used in deciding which clusters to merge. However, this does not result in a clustering that could be used to solve the K-means problem. Even though the local, perstep decisions try to minimize the squared error, the clusters produced on any level, except perhaps the very lowest levels, do not represent an optimal clustering from a "minimize global squared error" point of view. Furthermore, the clusters are not even "stable," in the sense that a point in a cluster may be closer to the centroid of some other cluster than to the centroid of its current cluster. However, Ward's method can be used as a robust method of initializing a Kmeans clustering. Thus, a local "minimize squared error" objective function seems to have some connection with a global "minimize squared error" objective function. Finally it is possible to attempt to fix up the hierarchical clustering produced by hierarchical clustering techniques. One idea is to move branches of the tree around so as to improve some global objective function. Another idea is to refine the clusters produced by a hierarchical technique by using an approach similar to that used for the multi-level refinement of graph partitions.

Limitations and Problems

We summarize the problems with agglomerative hierarchical clustering.

(1) No global objective function is being optimized.

(2) Merging decisions are final.

(3) Good local merging decisions may not result in good global results.

(4) Agglomerative hierarchical clustering techniques have trouble with one of more of the following: noise and outliers, non-convex shapes, and a tendency to break large clusters.

CURE

CURE (Clustering Using Representatives) is a clustering algorithm that uses a variety of different techniques to create an approach which can handle large data sets, outliers, and clusters with non-spherical shapes and non-uniform sizes. CURE represents a cluster by using multiple "representative" points from the cluster. These points will, in theory, capture the geometry and shape of the cluster. The first representative point is chosen to be the point furthest from the center of the cluster, while the remaining points are chosen so that they are farthest from all the previously chosen points. In this way, the representative points are naturally relatively well distributed. The number of points chosen, is a parameter, c, but it was found that a value of 10 or more worked well.

Once the representative points are chosen, they are shrunk toward the center by a factor, α. This helps moderate the effect of outliers, which are usually farther away from the center and thus, are "shrunk" more. For example, a representative point that was a distance of 10 units from the center would move by 3 units (for $\alpha = 0.7$), while a representative point at a distance of 1 unit would only move 0.3 units.

CURE uses an agglomerative hierarchical scheme to perform the actual clustering. The distance between two clusters is the minimum distance between any two 30 representative points (after they are shrunk toward their respective centers). While this scheme is not exactly like any other hierarchical scheme that we have seen, it is equivalent to centroid based hierarchical clustering if $\alpha = 0$, and roughly the same as single link hierarchical clustering if $\alpha = 1$. Notice that while a hierarchical clustering scheme is used, the goal of CURE is to find a given number of clusters as specified by the user.

CURE takes advantage of certain characteristics of the hierarchical clustering process to eliminate outliers at two different points in the clustering process. First, if a cluster is growing slowly, then this may mean that it consists mostly of outliers, since by definition outliers are far from others and will not be merged with other points very often. In CURE this first phase of outlier elimination typically occurs when the number of clusters is 1/3 the original number of points. The second phase of outlier elimination occurs when the number of clusters is on the order of K, the number of desired clusters. At this point small clusters are again eliminated.

Since the worst case complexity of CURE is $O(m^2 \log m)$, it cannot be applied directly to large data sets directly. For this reason, CURE uses two techniques for speeding up the clustering process. The first technique takes a random sample and performs hierarchical clustering on the sampled data points. This is followed by a final pass that assigns each remaining point in the data set to one of the clusters by choosing the cluster with closest representative point.

In some cases, the sample required for clustering is still too large and a second additional technique is required. In this situation, CURE partitions the sample data into p partitions of size m/p and then clusters the points in each partition until there are m/pq clusters in each partition. This pre-clustering step is then followed by a final clustering of the m/q intermediate clusters. Both clustering passes uses CURE's hierarchical clustering algorithm and are followed by a final pass that assigns each point in the data set to one of the clusters.

We summarize our description of CURE by explicitly listing the different steps:

(1) Draw a random sample from the data set. The CURE paper is notable for explicitly deriving a formula for what the size of this sample should be in order to guarantee, with high probability, that all clusters are represented by a minimum number of points.

(2) Partition the sample into p equal sized partitions.

(3) Cluster the points in each cluster using the hierarchical clustering algorithm to obtain m/pq clusters in each partition and a total of m/q clusters. Note that some outlier elimination occurs during this process.

(4) Eliminate outliers. This is the second phase of outlier elimination.

(5) Assign all data to the nearest cluster to obtain a complete clustering.

CHAMELEON

It is a new clustering algorithm that overcomes the limitations of existing agglomerative hierarchical clustering algorithm, CURE. CURE and different other related schemes emphasize cluster proximity while ignoring information regarding cluster interconnectivity. CHAMELEON operates on a sparse graph in which nodes represent data items, and weighted edges represent similarities among the data items. This sparse graph representation of the data set allows CHAMELEON to scale to large data sets and to operate successfully on data sets that are available only in similarity space and not in metric spaces. CHAMELEON finds the clusters in the data set by using a two phase algorithm. During the first phase, CHAMELEON uses a graph partitioning algorithm to cluster the data items into a large number of relatively small sub-clusters. During the second phase, it uses an agglomerative hierarchical clustering algorithm to find the genuine clusters by repeatedly combining together these sub-clusters. The key feature of CHAMELEON's agglomerative hierarchical clustering algorithm is that it determines the pair of most similar sub-clusters by taking into account both the inter-connectivity as well as the closeness of the clusters; and thus it overcomes the limitations of CURE that result from using only one of them.

Furthermore, CHAMELEON uses a novel approach to model the degree of inter-connectivity and closeness between each pair of clusters that takes into account the internal characteristics of the clusters themselves. Thus, it does not depend on a static user supplied model, and can automatically adapt to the internal characteristics of the clusters being merged. CHAMELEON's sparse graph representation of the data items is based on the commonly used *k*-nearest neighbor graph approach. Each vertex of the k-nearest neighbor graph represents a data item, and there exists an edge between two vertices, if data item corresponding to either of the nodes is among the *k*-most similar data points of the data point corresponding to the other node. Since CHAMELEON operates on a sparse graph, each cluster is nothing more than a sub-graph of the original sparse graph representation of the data set. There are several advantages of representing data using a k-nearest neighbor graph G.

- Firstly, data points that are far apart are completely disconnected in the Gk.
- Secondly, Gk captures the concept of neighborhood dynamically. The neighborhood radius of a data point is determined by the density of the region in which this data point resides.

5.7 DENSITY BASED METHODS

CLIQUE and MAFIA are density based clustering schemes, but because they are specifically designed for handling clusters in high-dimensional data, we will discuss DBSCAN and DENCLUE first.

DBSCAN

DBSCAN is a density based clustering algorithm that works with a number of different distance metrics. When DBSCAN has processed a set of data points, a point will either be in a cluster or will be classified as noise. DBSCAN is based on the concepts of a point being "density reachable" and "density connected." What follows is a more basic description. Conceptually, data points fall into three classes:

(*a*) **Core points.** These are points that are at the interior of a cluster. A point is an interior point if there are enough points in its neighborhood, *i.e.*, if the number of points within a given neighborhood around the point, as determined by the distance function and a supplied distance parameter, exceeds a certain threshold - also a supplied parameter. If two core points belong to each other's neighborhoods, then the core points belong to the same cluster.

(*b*) **Border points.** A border point is a point that is not a core point, *i.e.*, there are not enough points in its neighborhood, but it falls within the neighborhood of a core point. Note that a border point may fall within the neighborhoods of core points from several different clusters. If so, the cluster that the border point is grouped with will depend on the details of the DBSCAN algorithm, and is order dependent.

(*c*) **Noise points.** A noise point is any point that is not a core point or a border point.

Thus, for DBSCAN, a cluster is the set of all core points whose neighborhoods transitively connect them together, along with some border points.

DENCLUE

DENCLUE (DENsity CLUstEring) is a density clustering approach that takes a more formal approach to density based clustering by modeling the overall density of a set of points as the sum of "influence" functions associated with each point. The resulting overall density function will have local peaks, *i.e.*, local density maxima, and these local peaks can be used to define clusters in a straightforward way. Specifically, for each data point, a hill climbing procedure finds the nearest peak associated with that point, and the set of all data points associated with a particular peak (called a local density attractor) becomes a (center-defined) cluster. However, if the density at a local peak is too low, then the points in the associated cluster are classified as noise and discarded. Also, if a local peak can be connected to a second local peak by a path of data points, and the density at each point on the path is above a minimum density threshold, ξ, then the clusters associated with these local peaks are merged. Thus, clusters of any shape can be discovered.

DENCLUE is based on a well-developed area of statistics and pattern recognition which is known as "kernel density estimation". The goal of kernel density estimation (and many other statistical techniques as well) is to describe the distribution of the data by a function. For kernel density estimation, the contribution of each point to the overall density function is expressed by an "influence" (kernel) function. The overall density is then merely the sum of the influence functions associated with each point. Typically the influence or kernel function is symmetric (the same in all directions) and its value (contribution) decreases as the distance from the point increases. For example, for a particular point, x, the Gaussian function, $K(x) = e^{\frac{-\text{distance}(x,\, y)^2}{2\sigma^2}}$ is often used as a kernel function. (σ is a parameter which governs how quickly the influence of point drops off.) Figure 5.7 shows how a Gaussian function would look for a single two dimensional point, while figure 5.7(c) shows what the overall density function produced by the Gaussian influence functions of the set of points shown in figure 5.7(b).

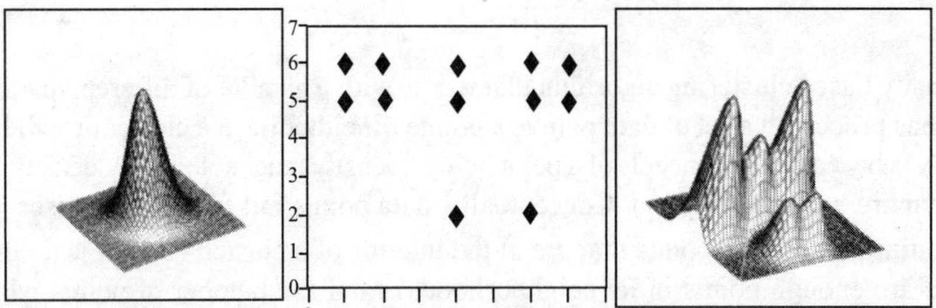

(a) Gaussian Kernel (b) Set of points (c) Overall density function

Figure 7. Example of the Gaussian influence (kernel) function and an overall density function. (σ = 0.75)

The DENCLUE algorithm has two steps, a preprocessing step and a clustering step. In the pre-clustering step, a grid for the data is created by dividing the minimal bounding hyper-rectangle into d-dimensional hyper-rectangles with edge length 2σ. The rectangles that contain points are then determined. (Actually, only the occupied hyperrectangles are constructed.) The hyper-rectangles are numbered with respect to a particular origin (at one edge of the bounding hyper-rectangle and these keys are stored in a search tree to provide efficient access in later processing. For each stored cell, the number of points, the sum of the points in the cell, and connections to neighboring population cubes are also stored.

For the clustering step DENCLUE, considers only the highly populated cubes and the cubes that are connected to them. For each point, x, the local density function is calculated only by considering those points that are from clusters which are (a) in clusters that are connected to the one containing the point and (b) have cluster centroids within a distance of $k\sigma$ of the point, where $k = 4$. As an efficiency measure, each point, x', on the path from x to its density attractor is assigned to the same cluster as x if dist $(x, x') \leq \sigma/2$.

As mentioned above, DENCLUDE discards clusters associated with a density attractor whose density is less than ξ. Finally, DENCLUE merges density attractors that can be joined by a path of points, all of which have a density greater than ξ. DENCLUE can be parameterized so that it behaves much like DBSCAN, but is much more efficient that DBSCAN. DENCLUE can also behave like K-means by choosing ξ appropriately and by omitting the step that merges center-defined clusters into arbitrary shaped clusters. Finally, by doing repeated clusterings for different values of σ, a hierarchical clustering can be obtained.

WaveCluster

WaveCluster is a clustering technique that interprets the original data as a two-dimensional signal and then applies signal processing techniques (the wavelet transform) to map the original data to a new space where cluster identification is more straightforward. More specifically, WaveCluster defines a uniform two-dimensional grid on the data and represents the points in each grid cell by the number of points. Thus, a collection of two-dimensional data points becomes an image, *i.e.*, a set of "gray-scale" pixels, and the problem of finding clusters becomes one of image segmentation. While there are a number of techniques for image segmentation, wavelets have a couple of features that make them an attractive choice. First, the wavelet approach naturally allows for a multiscale analysis, *i.e.*, the wavelet transform allows features, and hence, clusters, to be detected at different scales, *e.g.*, fine, medium, and coarse.

Secondly, the wavelet transform naturally lends itself to noise elimination. The basic algorithm of WaveCluster is as follows:

(1) Create a grid and assign each data object to a cell in the grid. The grid is uniform, but the grid size will vary for different scales of analysis. Each grid cell keeps track of the statistical properties of the points in that cell, but for wave clustering only the number of points in the cell is used.

(2) Transform the data to a new space by applying the wavelet transform. This results in 4 "subimages" at several different levels of resolutionan "average" image, an image that emphasizes the horizontal features, an image that emphasizes vertical features and an image that emphasizes corners.

(3) Find the connected components in the transformed space. The average subimage is used to find connected clusters, which are just groups of connected "pixels," *i.e.*, pixels which are connected to one another horizontally, vertically, or diagonally.

(4) Map the clusters labels of points in the transformed space back to points in the original space. WaveCluster creates a lookup table that associates each point in the original with a point in the transformed space. Assignment of cluster labels to the original points is then straightforward.

In summary, the key features of WaveCluster are order independence, no need to specify a number of clusters (although it is helpful to know this in order to figure out the right scale to look at, speed (linear), the elimination of noise and outliers, and the ability to find arbitrarily shaped clusters. While the WaveCluster approach can theoretically be extended to more than two dimensions, it seems unlikely that WaveCluster will work well (efficiently and effectively) for medium or high dimensions.

OPTICS

OPTICS ("Ordering Points To Identify the Clustering Structure") is an algorithm for finding density-based clusters in spatial data. It was presented by Mihael Ankerst, Markus M. Breunig, Hans-Peter Kriegel and Jörg Sander. Its basic idea is similar to DBSCAN, but it addresses one of DBSCAN"s major weaknesses: the problem of detecting meaningful clusters in data of varying density. In order to do so, the points of the database are (linearly) ordered such that points which are spatially closest become neighbors in the ordering. Additionally, a special distance is stored for each point that represents the density that needs to be accepted for a cluster in order to have both points belong to the same cluster. This is represented as a dendrogram. OPTICS requires two parameters: , which describes the maximum distance (radius) to consider, and , describing the number of points required to form a cluster. A point is a core point if at least points are found within its -neighborhood. Contrary to DBSCAN, OPTICS also considers points that are part of a more densely packed cluster, so each point is assigned a core distance that basically describes the distance to its th point:

$$\text{core-distance}_{\varepsilon,\text{ Min Pls}}(p) = \begin{cases} \text{UNDEFINED} & \text{if } |N_\varepsilon(p)| < \text{Min Pts} \\ \text{distance to the Min Pts the point} & \text{otherwise} \end{cases}$$

The reachability-distance of a point from another point is the distance between and, or the core distance of :

$$\text{reachability-distance}_{\varepsilon,\text{ Min Pls}}(p, 0) = \begin{cases} \text{UNDEFINED} & \text{if } |N_\varepsilon(0)| < \text{Min Pts} \\ \text{Max (core-distance}_\varepsilon(0)\text{, distance (0, p)} & \text{otherwise} \end{cases}$$

Basically, if and are nearest neighbours, this is the $\varepsilon' < \varepsilon$ we need to assume in order to have and belong to the same cluster.

Both the core-distance and the reachability-distance are undefined if no sufficiently dense cluster (w.r.t. ε) is available. Given a sufficiently large ε, this will never happen, but then every ε-neighborhood query will return the entire database, resulting in $O(n^2)$ runtime. Hence, the ε parameter is required to cut off the density of clusters that is no longer considered to be interesting and to speed up the algorithm this way.

The parameter ε is strictly speaking not necessary. It can be set to a maximum value. When a spatial index is available, it does however play a practical role when it comes to complexity. It is often claimed that OPTICS abstracts from DBSCAN by removing this parameter, at least to the amount of only having to give a maximum value.

The basic approach of OPTICS is similar to DBSCAN, but instead of maintaining a set of known, but so far unprocessed cluster members, a priority queue (*e.g.*, using an indexed heap) is used. The algorithm is as follows:

```
OPTICS(DB, eps, MinPts)
    for each point p of DB
        p.reachability-distance = UNDEFINED
    for each unprocessed point p of DB
        N = getNeighbors(p, eps)
        mark p as processed
        output p to the ordered list
        Seeds = empty priority queue
        if (core-distance(p, eps, Minpts) != UNDEFINED)
            update(N, p, Seeds, eps, Minpts)
            for each next q in Seeds
                N' = getNeighbors(q, eps)
                mark q as processed
                output q to the ordered list
                if (core-distance(q, eps, Minpts) != UNDEFINED)
                update(N', q, Seeds, eps, Minpts)
```

In update(), the priority queue Seeds is updated with the -neighborhood of and , respectively:

```
update(N, p, Seeds, eps, Minpts)
    coredist = core-distance(p, eps, MinPts)
    for each o in N
        if (o is not processed)
        new-reach-dist = max(coredist, dist(p,o))
        if (o.reachability-distance == UNDEFINED) // o is not in Seeds
            o.reachability-distance = new-reach-dist
            Seeds.insert(o, new-reach-dist)
    else // o in Seeds, check for improvement
        if (new-reach-dist < o.reachability-distance)
        o.reachability-distance = new-reach-dist
        Seeds.move-up(o, new-reach-dist)
```

OPTICS hence outputs the points in a particular ordering, annotated with their smallest reachability distance (in the original algorithm, the core distance is also exported, but this is not required for further processing).

5.8 GRID BASED METHODS—STING AND CLIQUE

MIN or Single Link

For the single link or MIN version of hierarchical clustering, the proximity of two clusters is defined to be minimum of the distance (maximum of the similarity) between any two points in the different clusters. (The technique is called single link because, if you start with all points as singleton clusters and add links between points, strongest links first, then these single links combine the points into clusters.) Single link is good at handling non-elliptical shapes, but is sensitive to noise and outliers. The following table gives a sample similarity matrix for five items (I1 – I5) and the dendogram shows the series of merges that result from using the single link technique.

	I1	I2	I3	I4	I5
I1	1.00	0.90	0.10	0.65	0.20
I2	0.90	1.00	0.70	0.60	0.50
I3	0.10	0.70	1.00	0.40	0.30
I4	0.65	0.60	0.40	1.00	0.80
I5	0.20	0.50	0.30	0.80	1.00

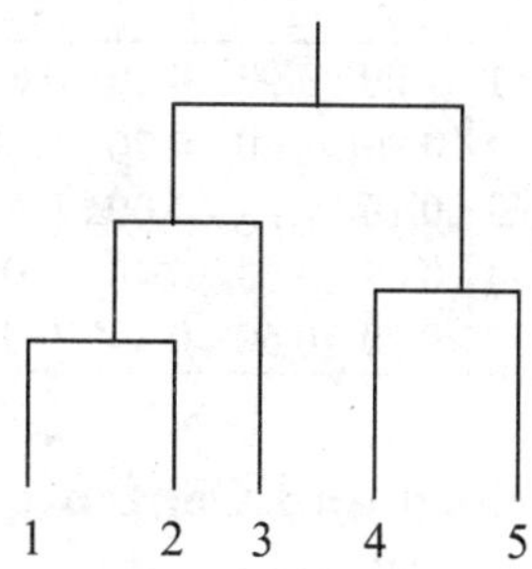

MAX or Complete Link or CLIQUE

For the complete link or MAX version of hierarchical clustering, the proximity of two clusters is defined to be maximum of the distance (minimum of the similarity) between any two points in the different clusters. (The technique is called complete link because, if you start with all points as singleton clusters, and add links between points, strongest links first, then a group of points is not a cluster until all the points in it are completely linked, *i.e.*, form a clique.) Complete link is less susceptible to noise and outliers, but can break large clusters, and has trouble with convex shapes. The following table gives a sample similarity matrix and the dendogram shows the series of merges that result from using the complete link technique.

	I1	I2	I3	I4	I5
I1	1.00	0.90	0.10	0.65	0.20
I2	0.90	1.00	0.70	0.60	0.50
I3	0.10	0.70	1.00	0.40	0.30
I4	0.65	0.60	0.40	1.00	0.80
I5	0.20	0.50	0.30	0.80	1.00

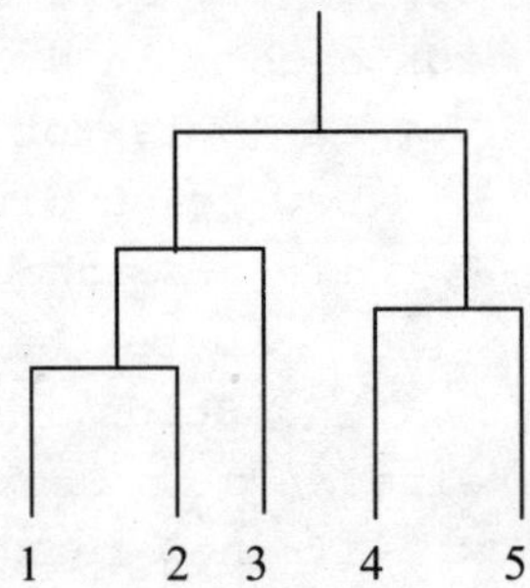

Group Average

For the group average version of hierarchical clustering, the proximity of two clusters is defined to be the average of the pairwise proximities between all pairs of points in the different clusters. Notice that this is an intermediate approach between MIN and MAX. This is expressed by the following equation:

$$\text{proximity (cluster 1, cluster 2)} = \frac{\sum\limits_{\substack{p_1 \,\varepsilon\, \text{cluster 1} \\ p_1 \,\varepsilon\, \text{cluster 2}}} \text{proximity } (p_1, p_2)}{\text{size (cluster 1)} * \text{size (cluster 2)}}$$

The following table gives a sample similarity matrix and the dendogram shows the series of merges that result from using the group average approach. The hierarchical clustering in this simple case is the same as produced by MIN.

	I1	I2	I3	I4	I5
I1	1.00	0.90	0.10	0.65	0.20
I2	0.90	1.00	0.70	0.60	0.50
I3	0.10	0.70	1.00	0.40	0.30
I4	0.65	0.60	0.40	1.00	0.80
I5	0.20	0.50	0.30	0.80	1.00

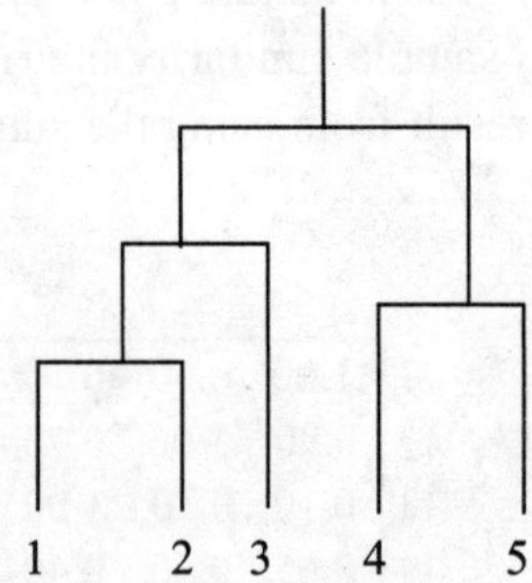

Ward's Method and Centroid Methods

For Ward's method the proximity between two clusters is defined as the increase in the squared error that results when two clusters are merged. Thus, this method uses the same objective function as is used by the K-means clustering. While it may seem that this makes this technique somewhat distinct from other hierarchical techniques, some algebra will show that this technique is very similar to the group average method when the proximity between two points is taken to be the square of the distance between them.

Centroid methods calculate the proximity between two clusters by calculating the distance between the centroids of clusters. These techniques may seem similar to Kmeans, but as we have remarked, Ward's method is the correct hierarchical analogue. Centroid methods also have a characteristic - often considered bad - that other hierarchical clustering techniques don't posses: the possibility of inversions. In other words, two clusters that are merged may be more similar than the pair of clusters that were merged in a previous step. For other methods the similarity of clusters merged decreases (the distance between merged clusters increases) as we proceed from singleton clusters to one all inclusive clusters.

STING

STING is a grid based multi resolution clustering technique in which the spatial area is divided into rectangular cells (using latitude and longitude) and employs a hierarchical structure. There are usually several levels of such rectangular cells corresponding to different levels of resolution. Each cell at a high level is partitioned to form child cells at lower level. A cell in level i corresponds to union of its children at level $i + 1$. Each cell (except the leaves) has 4 children and each child corresponds to one quadrant of the parent cell. Statistical information regarding the attributes in each grid cell (such as, mean, Standard Deviation maximum & minimum values) is pre computed and stored. Statistical parameters of higher level cells can easily be computed from the parameters of lower level cells. For each cell, there are attribute independent parameters and attribute dependant parameters:

(*i*) Attribute independent parameter: count

(*ii*) Attribute dependant parameters

When the data are loaded into the database, the parameters count, m, s, min, max of the bottom level cells are calculated directly from the data where,

- M-Mean of all values in this cell;
- S-Standard deviation of all values in this cell;
- Min-minimum value of the attribute in this cell;
- Max-minimum value of the attribute in this cell;
- Distribution-Type of distribution the attribute value follows. Distribution types are normal, uniform exponential and none.

Value of distribution may either be assigned by the user or obtained by hypothesis tests such as X2 test. First, a layer is determined from which the query processing process is to start. This layer may consist of small number of cells. For each cell in this layer we check the relevancy of cell by computing confidence internal. Irrelevant cells are removed and this process is repeated until the bottom layer is reached.

5.9 MODEL BASED METHOD

(*a*) **Statistical Approach:** Statistics is the traditional field that deals with the quantification, collection, analysis, interpretation, and drawing conclusions from data. Data mining is an interdisciplinary field that draws on computer sciences (data base, artificial intelligence, machine learning, graphical and visualization models), statistics and engineering (pattern recognition, neural networks). DM involves the analysis of large existing data bases in order to discover patterns and relationships in the data, and other findings (unexpected, surprising, and useful). Typically, it differs from traditional statistics on two issues: the size of the data set and the fact that the data were initially collected for purpose other than that of DM analysis. Thus, experimental design, a very important topic in traditional statistics, is usually irrelevant to DM. On the other hand asymptotic analysis, sometimes criticized in statistics as being irrelevant, becomes very relevant in DM. While in traditional statistics a data set of 100 to 104 entries is considered large, in DM even 104 may be considered a small set fit to be used as an example, rather than a problem encountered in practice. Problem sizes of 107 to 1010 are more typical. It is important to emphasize, though, that data set sizes are not all created equal. One needs to distinguish between the number of cases (observations) in a large data set (*n*), and the number of features (variables) available for each case (*m*). In a large data set, n, m or both can be large. Moreover these definitions may change when the same data set is being used for two different purposes. A nice demonstration of such an instance can be found in the 2001 KDD competition, where in one task the number of cases was the number of purchasing customers, the click information being a subset of the features, and in the other task the clicks were the cases.

(*b*) **Neural Network approach:** Neural network method is used for classification, clustering, feature mining, prediction and pattern recognition. It imitates the neurons structure of animals, bases on the M-P model and Hebb learning rule, so in essence it is a distributed matrix structure. Through training data mining, the neural network method gradually calculates (including repeated iteration or cumulative calculation) the weights the neural network computes. The neural network model can be broadly divided into the following three types:

1. **Feed-forward networks:** it regards the perception back-propagation model and the function network as representatives, and mainly used in the areas such as prediction and pattern recognition;
2. **Feedback network:** it regards Hopfield discrete model and continuous model as representatives, and mainly used for associative memory and optimization calculation;
3. **Self-organization networks:** it regards adaptive resonance theory (ART) model and Kohonen model as representatives, and mainly used for cluster analysis.

At present, the neural network most commonly used in data mining is BP network. Of course, artificial neural network is the developing science, and some theories have not really taken shape, such as the problems of convergence, stability, local minimum and parameters adjustment. For the BP network the frequent problems it encountered are that the training is slow, may fall into local minimum and it is difficult to determine training parameters. Aiming at these problems some people adopted the method of combining artificial neural networks and genetic gene algorithms and achieved better results. Artificial neural network has the characteristics of distributed information storage, parallel processing, information, reasoning, and self-organization learning, and has the capability of rapid fitting the non-linear data, so it can solve many problems which are difficult for other methods to solve.

(*c*) **Outlier Analysis:** Outliers can occur by chance in any distribution, but they are often indicative either of measurement error or that the population has a heavy-tailed distribution. In the former case one wishes to discard them or use statistics that are robust to outliers, while in the latter case they indicate that the distribution has high kurtosis and that one should be very cautious in using tools or intuitions that assume a normal distribution. A frequent cause of outliers is a mixture of two distributions, which may be two distinct sub-populations, or may indicate 'correct trial' versus 'measurement error'; this is modeled by a mixture model.

In larger samplings of data, some data points will be further away from the sample means than what is deemed is reasonable. This can be due to incidental systematic error or flaws in the theory that generated an assumed family of probability distributions, or it may be that some observations are far from the center of the data. Outlier points can therefore indicate faulty data, erroneous procedures, or areas where a certain theory might not be valid. However, in large samples, a small number of outliers is to be expected (and not due to any anomalous condition).

Outliers, being the most extreme observations, may include the sample maximum or sample minimum, or both, depending on whether they are extremely high or low. However, the sample maximum and minimum are not always outliers because they may not be unusually far from other observations.

Naive interpretation of statistics derived from data sets that include outliers may be misleading. For example, if one is calculating the average temperature of 10 objects in a room, and nine of them are between 20 and 25 degrees Celsius, but an oven is at 175 °C, the median of the data could be between 20 and 25 °C but the mean temperature will be between 35.5 and 40 °C. In this case, the median better reflects the temperature of a randomly sampled object than the mean; however, naively interpreting the mean as "a typical sample", equivalent to the median, is incorrect. As illustrated in this case, outliers may be indicative of data points that belong to a different population than the rest of the sample set. Estimators capable of coping with outliers are said to be robust: the median is a robust statistic, while the mean is not.

The choice of how to deal with an outlier should depend on the cause.

1. **Retention:** Even when a normal distribution model is appropriate to the data being analyzed, outliers are expected for large sample sizes and should not automatically be discarded if that is the case. The application should use a classification algorithm that is robust to outliers to model data with naturally occurring outlier points.
2. **Exclusion:** Deletion of outlier data is a controversial practice frowned on by many scientists and science instructors; while mathematical criteria provide an objective and quantitative method for data rejection, they do not make the practice more scientifically or methodologically sound, especially in small sets or where a normal distribution cannot be assumed. Rejection of outliers is more acceptable in areas of practice where the underlying model of the process being measured and the usual distribution of measurement error are confidently known. An outlier resulting from an instrument reading error may be excluded but it is desirable that the reading is at least verified.

   In regression problems, an alternative approach may be to only exclude points which exhibit a large degree of influence on the parameters, using a measure such as Cook's distance. If a data point (or points) is excluded from the data analysis, this should be clearly stated on any subsequent report.
3. **Non-normal distributions:** The possibility should be considered that the underlying distribution of the data is not approximately normal, having "fat tails". For instance, when sampling from a Cauchy distribution, the sample variance increases with the sample size, the sample mean fails to converge as the sample size increases, and outliers are expected at far larger rates than for a normal distribution.
4. **Alternative models:** In cases where the cause of the outliers is known, it may be possible to incorporate this effect into the model structure, for example by using a hierarchical Bayes model or a mixture model

5.10 VALIDATION OF CLUSTER ANALYSIS METHODS

Cluster analysis is a complicated interactive process, which makes it impossible to provide an entirely clear-cut prescription on how to do clustering or to perform cluster validation. In general, the experimental set-up should fundamentally differ depending on the primary aim of a study. Cluster validation aimed at the evaluation of a novel algorithm or the comparison of several algorithms should be quite different to the type of cluster validation used during the analysis of a novel biological dataset. This section attempts to give some general guidelines on the conduct of an effective cluster validation in both scenarios.

Cluster Validation for The Evaluation/Comparison of Algorithms

When evaluating algorithms, the choice of datasets is a primary issue. Certainly, several datasets should be used, not just one especially not only the dataset the algorithm was initially developed on. It is fundamental to appreciate that algorithms make different assumptions about the cluster structures, and are, consequently, more or less suited for particular datasets: no single algorithm can therefore be expected to perform well for all types of data. Thus, the aim of any evaluation study should not be to show that a particular algorithm is the best overall, but to show what the particular strengths and weaknesses of a given algorithm are. For this purpose, it is important to test on benchmarks with interesting known data properties. In this scenario, two types of questions are then of interest, which are both essential to understand fully the outcome of an experiment.

- *How well does the algorithm perform on a given dataset?* On benchmarks, this type of question can be objectively answered using external cluster validation. The use of adjusted validity measures is preferable.
- *Why is the algorithm not performing well? What is going wrong?* Internal validation technique can be used to highlight these issues, particularly those of Type 1, Type 2 and Type 3, and their combination in Pareto plots, as these have straightforward interpretations in terms of data properties.

Cluster Validation for a Novel Dataset

When clustering a novel biological dataset, cluster validation plays a very different role. A completely objective validation of cluster quality is usually impossible in such a case, but the use of cluster validation at different steps during the clustering process can help to improve the quality of results, and increase the confidence in the final result. Cluster analysis usually involves a first exploratory step, where the data are visualized (projected) to two- or three-dimensions (using methods such as principal components analysis or multi-dimensional scaling) in order to check for clustering tendencies. At this stage, a statistical test of clustering tendency may help to quantify the visual impressions obtained.

No entirely reliable method exists to identify the number of clusters in a dataset, and the choice of the best number of clusters may well depend on the clustering method used. A cluster analysis should therefore always be performed for a (sensible) range of different numbers of clusters. Access to such a sequence of solutions is essential to understand the operation of a clustering algorithm and to identify trends in the data.

The core cluster analysis should preferably be conducted using several conceptually distinct clustering algorithms, *i.e.* algorithms that are not biased towards the same type of clusters. Binary external indices can then be used to quantify analytically the similarity between clustering results (including those with different numbers of clusters). If conceptually different algorithms generate highly similar partitions, this is a good indicator that actual structure has been discovered.

On the other hand, coinciding clustering results returned by k-means, partitioning-around medoids or SOMs are less significant, as these algorithms share many concepts. If the partitions generated by different algorithms are highly dissimilar this is often an indication of poor structure in the data, and may point to defects in the pre-processing. In high-dimensional biological data, the structures in the data cannot often be perceived in the full feature space, and a drastic reduction of variables may be necessary in order to reduce the impact of noise. This process of feature selection is often necessary but should preferably be based on unsupervised methods (*e.g.* by selecting the variables with the highest variation across the dataset). If the features are selected using the knowledge of the real class labels (*e.g.* by selecting the variables which are best correlated with the known class structure), a subsequent cluster analysis will trivially yield the desired result (even for random data).

Internal validation measures should be used in addition to the above to provide feedback on the quality of the data and to check whether a given partitioning is justified in terms of the underlying data distribution. Here, it is important to use measures of the different basic types, Type 1, Type 2 and Type 3, and to check how well the solutions perform under each of them. A good clustering solution tends to perform reasonably well under multiple measures. If a solution performs well only under one of them, this is likely to be an artefact of the biases of the employed algorithm. Type 4 measures and plots in two-objective space may be a valuable tool in identifying solutions that perform consistently well. Given the noisy nature of biological data, robust measures like the Silhouette Width are generally preferable to noise-sensitive measures such as the Dunn Index.

Owing to the many sources of noise and the high dimensionality of the data, the above internal validation techniques on their own may often be insufficient in biological data analysis. Frequently, the most conspicuous structure in the data may be artefacts due to experimental factors. On the one hand, cluster analysis can be a valuable tool in identifying such artefacts. On the other hand, the artefacts will ultimately have to be removed if a researcher is interested in biologically meaningful results. Towards this goal, external unary measures can be applied to assess the degree of preservation of replicate-relationships, or of prior biological knowledge. This information can then provide additional feedback on the quality of the data and of previous pre-processing steps. A good final clustering result will ideally combine validity under both internal and external measures, *i.e.*, it will exhibit a distinct underlying cluster structure while being consistent with prior biological knowledge.

5.11 APPLICATIONS OF CLUSTERING TECHNIQUES IN ENGINEERING

Clustering techniques can be useful tools for exploring the underlying structure of a given data set and are being applied in a wide variety of engineering disciplines. This section discusses some applications of clustering techniques in engineering. Different clustering algorithms implicitly have different biases and therefore often have their own strengths when handling data sets with particular properties. It can be argued that there will not be a universal clustering algorithm that could be applied to arbitrary data. It is therefore important to have a clear understanding of the requirements of a particular application and to choose the technique that best fits those requirements. In general, to be useful in an engineering application, a clustering algorithm should have the following abilities:

- dealing with different types of data (numerical, categorical, text and images),
- handling noise, outliers and fuzzy data,
- discovering clusters of irregular shapes,
- dealing with large data sets and data of high dimensions,
- producing results that are easy to understand,
- being insensitive to the order of the input data,
- being simple to implement.

There are several engineering areas where clustering has been successfully employed as an essential step.

SUMMARY

- Cluster analysis is an exploratory data analysis tool which aims at sorting different objects into groups in a way that the degree of association between two objects is maximal if they belong to the same group and minimal otherwise.
- The purpose of cluster analysis is to discover a system of organizing observations, usually people, into groups where members of the groups share properties in common.
- Hierarchical clustering methods do not require preset knowledge of the number of groups.
- Two general methods of hierarchical clustering methods are available: divisive and agglormerative.
- Hierarchical techniques produce a nested sequence of partitions, with a single, all inclusive cluster at the top and singleton clusters of individual points at the bottom.
- The objective of K-medoid clustering is to find a non-overlapping set of clusters such that each cluster has a most representative point
- CLARANS was one of the first clustering algorithms that was developed specifically for use in spatial data mining.
- CLARANS uses a randomized search approach to improve on both CLARA and PAM.
- In hierarchical clustering the goal is to produce a hierarchical series of nested clusters, ranging from clusters of individual points at the bottom to an all-inclusive cluster at the top.
- Centroid methods calculate the proximity between two clusters by calculating the distance between the centroids of clusters.
- CURE (Clustering Using Representatives) is a clustering algorithm that uses a variety of different techniques to create an approach which can handle large data sets, outliers, and clusters with non-spherical shapes and non-uniform sizes.
- CURE represents a cluster by using multiple "representative" points from the cluster.
- DBSCAN is based on the concepts of a point being "density reachable" and "density connected."

REVIEW QUESTIONS

1. What do you understand by the term clustering? What are the different requirements in clustering?
2. Give some of the areas where clustering is used these days.
3. Describe in detail various categories of clustering methods.
4. Discuss k-means algorithm with the help of an example. Compare its features with k-medoid.
5. How partitional algorithms work with the help of minimum spanning tree approach?
6. What is meant by hierarchical clustering? Explain it with the help of a suitable example.
7. Write short notes on;
   (*a*) Single link method
   (*b*) Complete Link method
   (*c*) Group Average Link Method

■■■

CHAPTER 6

Introduction to Data Warehousing

These days there has been a great increase in processing capabilities and complexity of database management system which has made advancement in the field known as data warehousing. Data warehousing is a field that has evolved in the recent years and now which has become an important aspect of almost all type of industries depending upon online transactions. Data warehousing increases the capabilities of databases by providing additional space and functionality. In almost every database management system there are many users of the system. Different users are authorized to view data at different levels. Some users may only need to read some amount of data. Now this data is presented to users from multiple databases. Now it requires the people in the management to view the summarized data to make some decisions for the company. This is provided by data mining, data warehousing and OLAP (Online Analytical Processing). In this chapter we will focus our study to the concepts of data warehousing.

In section 6.1 we will study about some basics of data warehouse, data warehouse features, etc. Section 6.2 describes delivery process of data warehouse. Section 6.3 describes difference between operation database and a warehouse. In Section 6.4 we will study about components of warehouse. Section 6.5 describes dimensional modeling and various schemas. Section 6.6, 6.7 focuses on data marting and data warehouse architecture and its implementation respectively. Section 6.8 focuses on 3 tier architecture of data warehouses. In Section 6.11 we will study about metadata.

6.1 DATA WAREHOUSE

Today data warehouses are an important part of modern companies. For example any travel agent company like 'makemytrip.com' requires some data from past that at some time of year like in winter how many people like to travel Shimla for vacation. They will use this information or data from past to plan their strategies in the current year. However some additional information like snowfall in the current year also play its part but the number of train and plane tickets booked last year and number of hotels booked last year is always critical. This information from past gives an idea to the firm and hence it can plan its future actions. The main and ultimate goal of data warehouse is to record and analyze data from past. It is not designed for online transactions. Data warehouse is summarized information from various sources.

The formal definition of data warehouse is given by W.H. Inmon:

"A data warehouse is a subject-oriented, integrated, time-variant, non-volatile collection of data in support of management decisions".

When making a data warehouse, data from various sources (compatible or incompatible) is considered and converted to desired form. This data is required and analyzed by business professionals. As these people are not technical, they are only interested in simple and summarized information like list of 'A' grade customers or list of products in high demand.

6.1.1 Data Warehouse Features

In the above definition of data warehouse given by W.H. Inmon, he has described data warehouse as being subject-oriented, integrated, time variant and volatile. Now we are about to study what these terms actually mean.

The features of data warehouse are as follows:

1. **Subject Oriented:** The data warehouses are on particular subjects like sale of a product in last few months or production of a product in manufacturing department, etc. Data warehouses are made to analyze the historic data.
2. **Integrated:** The data warehouse on a particular subject is not from a single source. It is obtained from various compatible and incompatible sources and is integrated and is resolved for any anomalies and conflicts.
3. **Non Volatile:** The data warehouse is non volatile. By non volatile we mean that once the data is entered in it is not changed. It can be seen that data warehouse is the data from past that has occurred and is used for analytical purposes. Since it is not used for any online processing, there is no question of recovery and concurrency control techniques being described in database management system. Hence data in a data warehouse is non volatile.
4. **Time Variant:** We have already seen that data in a data warehouse is from past hence it depends variably on time. For example in a business firm, the people in the marketing department may be required to analyze the change in demand of their product over last few weeks.

6.1.2 Why do we Require Data Warehouse

BUSINESS USERS
STORE HISTORIC DATA
SELECTIVE DATA
DIFFERENTIALE ANALYTICAL AND OPERATIONAL PROCESSING
MAKE STRATEGIC DECISIONS

Figure 6.1. Need of Data Warehouse

Business Users: Business users require data warehouse to view summarized data from past. Since these people are non technical, the data may be presented to them in a very simple form such that it is very easy to understand the facts and figures.

Store Historic Data: Data warehouse is required to store the time variable data from past. Hence data warehouse is time variable. This data is made to be used for various purposes.

Selective Data: When data is stored in data warehouse it may not be full data. This is because data warehouse contains the summarized data. The other reason is that various users are entitled to view only a portion of data and not the complete data.

Differentiate Analytical and Operational Processing: The key requirement that separates operational database from data warehouse is that operational database is entitled for online transactions and various operations. Now if we make this database for analytical purposes then it causes an extra burden or load on operational database. Hence we require an additional database that gives the requisite data in summarized form and which can be used for analysis. This reduces the load on operational database.

Make Strategic Decisions: Some strategies may be depending upon the data in data warehouse. Hence data warehouses contribute in making strategic decisions.

6.2 DELIVERY PROCESS: BUILDING A DATA WAREHOUSE

So far we have studied about what data warehouse really is. Now in this section we are about to discuss the delivery process of data warehouse. Before we start further discussion it is necessary that we understand that what is a delivery process? In software environments every software project or assignment has a delivery process. In software engineering we have studied about waterfall model. It is the simplest model for any project. But the delivery process that delivers the data warehouse has to be logically different from the waterfall model. The reason behind this is that it is difficult to complete the tasks and deliverables in a strict environment by waterfall model. This is because the requirements are not properly understood and may change over the time in future.

6.2.1 Steps in Data Warehouse Delivery Process

1. **IT Strategy:** The first step in the delivery process of data warehouse is IT Strategy. Data warehouses are strategic investments. It requires business process to redesign in order to generate projected benefits. In case if there is no overall strategy for data warehouse then the risks are increased.
2. **Business Case:** After the IT Strategy has been designed the next step is the business case. It involves identification of projected benefits. The projected benefit cannot be laid in terms of figures (not quantifiable). The benefits that are projected should be clear. The data warehouses that do not have business case involve risk of suffering from problems in the later stages of the delivery process.
3. **Education and Prototyping:** Sometimes companies and firms experiment on the basis of analysis from data obtained from warehouses. In some cases implementing data warehousing solutions is the first solution implemented by large companies. This requires solid technical skills, experience and hardware. Implementing prototyping on small scale may further delay the education process unless objectives have been properly understood and a feasibility study has been done and activity time scale is not critical.
4. **Business Requirements:** The next step in the delivery process is understanding business requirements. It is a very critical phase because if business requirements are not properly understood then it causes a data warehouse that is out of context. 20% of the time in business requirements should be spent on understanding long term requirements because it is evident that requirements tend to change with time.

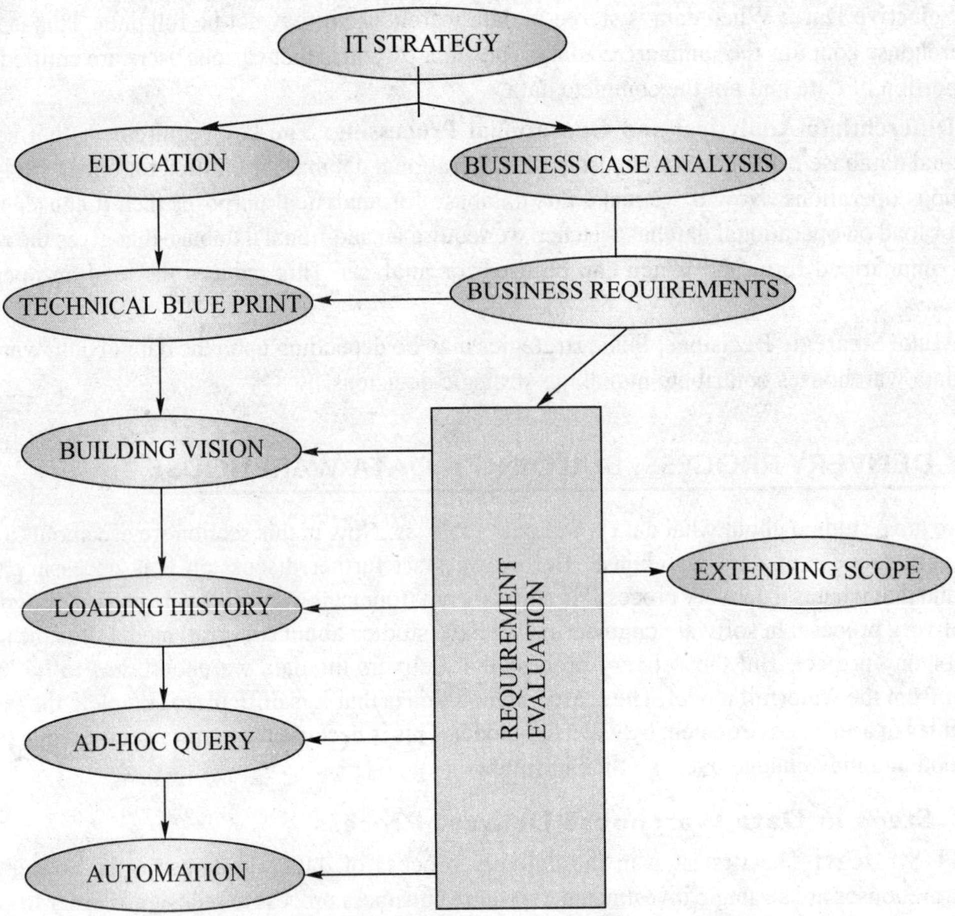

Figure 6.2. Steps in delivery process

5. **Technical Blue Print:** Technical blue print is an important step in the overall delivery process. This is because it lays architecture of warehouse. The concept of blue print is similar to the blue print that is drawn by an architect when building a building. Architect of building lays a plan that gives the number of floors, windows, their position, lift position, stairs and hundreds of other parameters. Similarly technical blue print of delivery process makes an architecture plan that satisfies long term requirements. It lays server and data mart architecture and essential components of the database design. It makes a strategy for data retention. Capacity plan for hardware and infrastructure is also required. Significant components are also identified.
6. **Building the vision:** In this step of delivery process the first production deliverable is produced. It is the smallest component of data warehouse that adds business benefit. It includes the components required for extracting and loading data. Load the data from one or two sources that includes lesser history.
7. **History Load:** The remaining history that was not loaded in the previous step is loaded in this state. This state also includes addition of new entities. The new physical entities created provide for storage of increased data volumes. Suppose a travel agent company loads data from last 2-3 months. This data would be sufficient to plan short term strategy. In history load it would be

required to load data from last 3-4 years which would be required to plan seasonal and long term strategies.

8. **Ad-hoc query:** In this phase ad-hoc query tool is configured to operate against Data Warehouse. These are the end user tools that have the capabilities to answer any question on a particular subject. Suppose a bank manager enters a question to display number of loans approved last month. This question is converted to database query by access tool which then displays the requisite data from data warehouse.
9. **Automation:** Data warehouse contains the warehouse with fully automated operational management process. It includes storing and loading the data from various sources. This data is then transformed to a suitable form to be analyzed. The data is backed up and restored. The query profile is monitored and appropriate aggregation for maintaining system performance is determined.
10. **Extending Scope:** This is the second last state of delivery process. It supports for extension of new business requirements and supports loading of new data from other data sources.
11. **Requirement Evolution:** This is the last stage of delivery process of data warehouse. As we all know that requirements are not static and evolve continuously. As the business requirements will change it supports to be reflected in the system.

This marks the end of delivery process. However since the requirements change over the time it is required to support these changes and be reflected in the system.

6.3 DIFFERENCE BETWEEN DATA WAREHOUSE AND DATABASE

So far we have discussed some basic concepts of data warehouse. We have also done some discussion that differentiates data warehouse from database. Basically a database contains complete data. However a warehouse contains smaller amounts of data that too in summarized form. The difference is evident because a database is used for operational purposes that support transactions. However the need of data warehouse has arose because it is required for analytical purposes. This reduces the additional load on operational databases.

The data warehouses and databases can be distinguished on the basis of following grounds:

1. **Orientation:** Databases are application oriented whereas data warehouses are subject oriented that is they are on particular subject.
2. **Amount of Details:** Databases contain a detailed data whereas warehouses only contain summarized data.
3. **Time Dependence:** Databases gives the data at the moment of access. Warehouses on the other hand give data over time.
4. **Community served:** Databases serve clerical community. This is evident from the fact that clerks mostly perform operations on data in the database. They perform online transactions. On the other hand data warehouses serve the managerial community because managers mostly play the role of analysis.
5. **Requirements:** Requirements are totally understood for processing before initial development in databases whereas requirements in data warehouses are not understood prior to its development.
6. **Volatility:** Operational data is volatile whereas warehouses are non volatile.

7. **Compatibility:** Operational databases are software life cycle driven whereas data warehouses follow completely different life cycle.
8. **Performance:** Databases are performance sensitive. Data warehouses are performance relaxed.
9. **Availability:** Databases support high availability whereas data warehouses support relaxed availability.
10. **Redundancy:** Operational data is usually non redundant whereas data warehouses usually contain some redundancies.

6.4 COMPONENTS OF DATA WAREHOUSES

There are three components of data warehouses:

1. Identifying the data source
2. Cleaning up
3. Transformation and migration tools

The three components namely identifying the data source, cleaning up and transformation and migration tools are responsible to carry out all the necessary actions that are to be required to extract the useful information from compatible or incompatible data sources, merging and splitting data from various data sources, condensation and summarization of data to transform the data to summarized form that can be used for analytical purposes. The functionalities performed by various components of data warehouse are as follows.

1. Leaving the data that is not required.
2. Summarizing the data.
3. Defining common data names and functionalities.
4. Reflecting the data in case data in source databases changes.

So far we have seen various components of data warehouses. Now let us discuss and try to understand functionality and role played by each of them one by one.

1. **Identifying the data source:** The first component of data warehouse is identifying the data source. Before starting this component the requirements of data warehouse are laid down. Some additional requirements that are expected to arise in the future are also laid down. After the requirements are clear in minds, the data sources from where the data is extracted are identified. These data sources may be some operational databases or some spreadsheets. The maximum data comes from 2-3 data sources. The requirements play their role as to which data sources are identified. To start in the beginning we consider the data source from which the maximum data comes. We then built around this source by considering additional data from other data sources. If the data from data source is to be used in data marts then the data is aggregated on the basis of level of granularity. Granularity is the information at the lowest level. Mostly most of the data required is identified from almost two data sources. The situation may arise when the data from some source may not be taken because the data is incompatible with the desired format of data.
2. **Cleaning up the data:** This is the second component of any warehouse. Data cleaning has three components.

   (*a*) Data validation

   (*b*) Enhancing data

(*c*) Handling errors

Data validation is entitled to perform a series of checks like if the value for a particular attribute is validated, the value of attribute is consistent with the tupples where this attribute occurs and with other tupples, attributes validated in context with same type of rows in other tables. It also validates relation between tupples in various tables.

Data enhancement is entitled to enhance the data by removing some anomalies and redundancies and improving its meaning. This can be understood from the following example in an employee system various departments like Human Resource Department, Accounts department maintain the records of various employees. This data is likely to become changed and become non-synchronized. Data enhancement plays the role of synchronization of similar data.

Finally error handling is responsible to handle the various errors that are likely to persist.

3. **Transformation and Migration Tools:** Transformation tools are required to transform the data from incompatible sources to requisite form. It includes managing attributes to convert it to an appropriate form. The different types of transformations that are performed are as follows.

   (*a*) Splitting fields

   (*b*) Consolidating fields

   (*c*) Standardizing

   (*d*) De-duplication

In splitting fields it may be required to split the source field into more than one target fields. For example name field could be split to three fields, first name, middle name and last name. Henry Jonathan Thomas under single field name could be split into three fields as follows.

First name: Henry

Middle Name: Jonathan

Last name: Thomas

In consolidating fields the data from two sources is consolidated into single target. For example,

```
First source: emp_email: abc@gmail.com
Second source: emp_email: abc@ymail.in
```

The above fields could be consolidated as follows:

```
Emp_email: abc@gmail.com abc@ymail.in
```

Some standardization is followed in storing data. For example date to be stored in MM- DD-YYYY format.

When data is stored in more than one space it is likely that some duplicity may occur. De-duplication is entitled to mange duplicate data. For example:

```
First source: Name: Henry J. Thomas
Second source: Name: Henry Jonathan Thomas
```

This information could be stored as

Name: Henry Jonathan Thomas

6.5 DIMENSIONAL MODELLING

The common aspects of business are conceptualized and visualized by techniques described by dimensional modelling. There are two concepts of dimensional modelling.

1. Facts:

(*a*) Facts consist of related data items and measures.

(*b*) Fact focuses on decision making process.

(*c*) Facts are described by valued attributes called measures.

(*d*) Fact is a measure of business

2. Dimension:

(*a*) Dimensions are parameters over which analysis of facts is performed.

(*b*) Dimension is a parameter that analyzes for example the total sale which is a fact.

Dimensional modelling is an important architecture for building data warehouses. Our data warehouse may involve multiple parameters giving rise to multi dimensional warehouses consisting of three or more dimensions. Summarized data is viewed by categories called dimensions. Cubes consist of processing units made of facts and dimensions

6.5.1 Multidimensional Data Models

Multidimensional models have emerged greatly in the last 30 years. Multi dimensional models are used when the main aim is to perform analysis of data rather than performing online transactions on data. Analysis on data is the main objective of developing data warehouses. Multidimensional database technology is used when it is required to analyze huge amount of data for making decisions. It is based on relational elements.

Multidimensional models are described by three main parameters:

1. Modeling business rules
2. Measures and cube
3. Dimensions

Data cube uses a technique called pivoting or rotation to change dimensional hierarchies. The data cube can be rotated and aligned with different axes. It is used to provide business modeling to the analysis. Measures, cubes and dimensions create a strong environment by mutual interaction. At each level new objects are added that increases interaction. Multi dimensional can be extended to support additional interaction between objects. It requires additional calculations from users to add the capability of isolating it from corresponding cube structure. Mostly multi dimensional database systems come under two categories.

1. Relational database systems make multi dimensional schemas like snowflake and star schema.
2. Some multi dimensional database systems are designed only for online purposes.

Factual table remains in contact with all other dimensional tables. Dimensional table is not connected with other dimensional tables. Relational database systems support online operations or transactions but do not support analytical processing used for decision purposes whereas multi dimensional database supports data warehouses and online transactions. Through a multi dimensional database a user can ask any question of subject and it is answered correctly. Today there are a number of multi dimensional database systems available in market which differ in data storage and interface. Multi dimensional database systems enhance the capabilities by supporting multi dimensional requests from users. It enables non technical people like analysts, managers to analyze performance of an enterprise.

Goals of multi dimensional data models: The goals of multi dimensional data models are to support analysis in a simple and faster way by executives, managers and business professionals. These people are not interested in the overall architecture. They are interested in the analysis of data. The online analytical processing engine supports for its physical implementation.

Application areas:

1. Multi dimensional data modeling is used in data warehousing where data from various data sources is collected to support for online analytical processing in a faster way for decision making strategies.
2. It is used in building online analytical processing engines that answers queries by merging data from various sources. Hence factual table should be implemented in a simpler manner.

6.5.2 Logical Multidimensional Data Model

Multidimensional data model is very simple. It is the simplicity of the multi dimensional data model that makes its users to completely understand databases very clearly and navigate between the databases. The multi dimensional data model contains dimensions, attributes, logical cubes, measures, hierarchies and levels. It is simple to understand and deploy because it views business level entities as objects. The data analysts know which levels, hierarchies are used, what needs to be examined and which dimensions and attributes should be considered. The basic model for online analysis is multi dimensional data cubes.

Given below in figure 6.3 is a basic diagram which describes relationship between dimensions, attributes, cube, measures and hierarchies. The observation tells us that cube cannot exist on its own. Logical cubes organize measures that have same dimensions. Relationships between measures of same cube have same relations with other logical objects. These are examined and written together. Various measures including time dimensions organize measures. Here we are performing multidimensional analysis which enables users to interact with each dimension separately. Queries with dimensions consist of a number of steps. Whereas relational model allow users to manipulate all the elements as a whole which tends to increase the probability of errors, multidimensional model allow users to use each dimension separately and is also user friendly. Multi step queries can be easily created. The sophisticated query is broken down into simpler and manageable queries. The relational implementation of multi dimensional model uses star or snowflake schema.

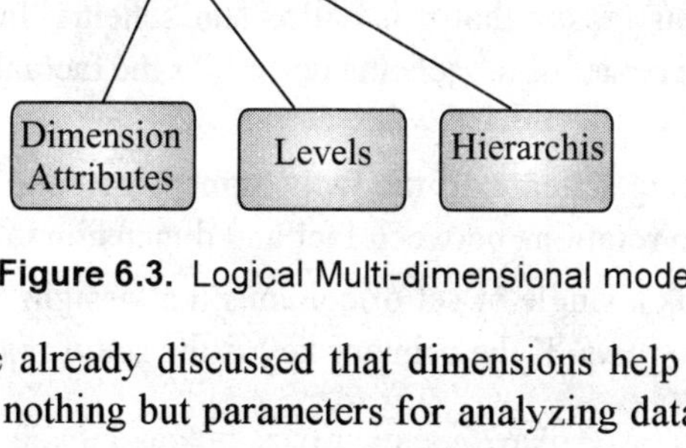

Figure 6.3. Logical Multi-dimensional model

Dimensions: We have already discussed that dimensions help to analyze data from various perspectives. Dimensions are nothing but parameters for analyzing data. These perspectives could be any like time. On the basis of time for example firm would like to analyze its turnover over the last five years. This analysis is very comfortable for any business employee.

Dimension tables contain dimension keys, attributes and values. Dimension employee will contain all the details working in a company, employee name, address, email, phone number, date of birth, etc. Similarly dimension customer will contain all the necessary details of all the customers of the company. The number of rows of the dimension table range from few hundreds to millions of rows. For example a company can have a thousand customers and hence thousand rows, one for each customer.

Employee dimension will be shown as follows:

```
Employee_key
Employee_first_name
Employee_middle_name
Employee_last_name
Employee_designation
Employee_city
```

However the employee database may contain other attributes too but when considering data warehouse it would be impractical to contain employee email address because it is not required for analysis. The multi dimensional table will also contain time dimension because when required for analysis it will be compared from time dimension that what is the progress or if the total revenue from sale of a product in product dimension has decreased. The department of a company may be present in more than location and hence will also contain location dimension. There are some mathematical rules that are summarized in a hyper cube. Example would be like revenue from different products would be summarized to a single product group.

Measures: Measures are stored in fact tables. The structure of dimension tables is hierarchical of attributes of quality. For connecting dimension tables and fact tables the primary key of dimension table is stored in fact table. The dimension hierarchies are modeled in the same dimension table. The dimension shall not be represented by a single table otherwise it leads to anomalies which are then required to be normalized. Measures are facts in numbers that have occurred in past like number of hours of work, total revenue, number of customers last month. There are three types of measures. Additive measures and semi additive measures. Additive measures are measures that can be added across all dimensions whereas semi additive measures are measures that cannot be added across all dimensions. Non additive measures cannot be added.

6.5.3 Star Schema

Star schema is the schema for data warehouse. It is the simplest and easiest to implement of all the schemas. Its shape is similar to star. The entity relationship diagram between fact table and dimension tables resembles a star. It is this reason that it is called star schema. In start schema some points are coming out from the center. The center of the schema resembles the factual tables and the points resemble the dimension tables.

First of all before creating schema all the facts, dimensions and desired level of granularity is considered. Then we define the relations between fact and dimension tables.

Primary key in a table is a single or set of columns that assigns unique value to every row and distinguishes it from rest of the rows. If the primary key is the set of many columns then it is called as composite key.

A combination of integers generates a synthetic key. Synthetic keys are system generated keys. Synthetic keys provide additional benefits over natural keys. The synthetic key is included in the fact table which needs to be joined with dimensional table.

Synthetic keys are more beneficial than natural keys in the way that natural keys are a combination of string whereas synthetic keys are combination of integers. In this way response time to synthetic keys is improved.

When the data from database is required to be translated to star schema it requires proper mapping. This mapping may require many transformations like splitting fields, de-duplication, aggregation. We have already seen that star schema is the simplest and the easiest to implement.

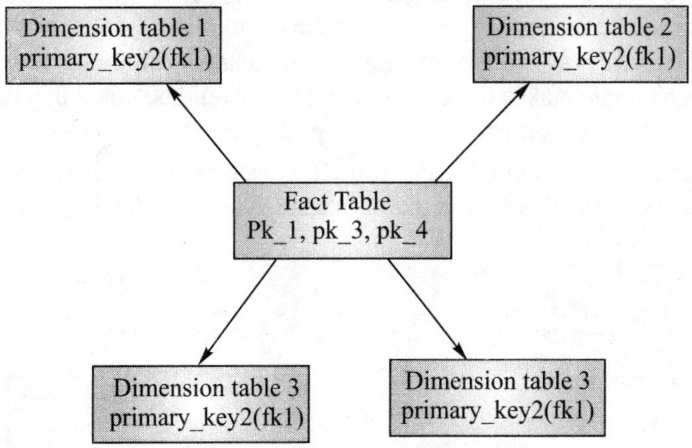

Figure 6.4. Star Schema

Fact tables are large whereas dimension tables are smaller. The fact tables contain primary key attributes. The dimension tables contain information about entries corresponding to an attribute in the fact table.

A star query joins fact table to a dimension table. As we have discussed that fact table contains a foreign key from dimension table that joins fact table to the dimension table. The dimension tables are not joined to one another. Whenever a star query is encountered the optimizer identifies it and creates an execution plan for it.

The star schema has many advantages. The business requirements by users are identified as entities and are mapped in the star schema. Whenever a star query is encountered the optimizer identifies it and creates an execution plan for it. Star schema is suitable for small data warehouses as well as large data warehouses.

6.5.4 Snowflake Schema

Snowflake schema is designed by transforming star schema to snowflake schema. The star schema is not normalized. This is due to presence of many anomalies and redundancies in the dimension tables. These dimension tables are then normalized to third normal form. The dimension table is a large table. Not if we want to normalize this table then we need to divide this dimension table to a number of dimension tables then we need a number of primary key attributes that are included as foreign key attributes so that the dimension tables are joined.

This conversion of star schema to snowflake schema reduces the space. But the problem that arises is that it makes the schema sophisticated than star schema. Hence the efficiency and performance of snowflake queries decreases.

We have already studied that star schema is very simple. But it suffers from some anomalies and is not normalized. Hence we convert it to a complex but normalized schema called snowflake schema. This however reduces the performance of queries.

Let us now try to explain this with an example. Consider the university management system. Consider the entity named course_offering. Now this course_offering table is included in the dimension table. This is a single dimension table containing no join. So it takes more space. This however gives performance to the queries due to its simplicity. But course_offering is not normalized and hence contains many anomalies. So it requires to be normalized. The course_offering dimension table is divided into two dimension tables joined on course_id. The two dimension tables are course and professor that are joined on course_id which is a primary attribute of dimension table course used as foreign key in professor table. This example makes the situation clear. It should be noted that division into two tables has however reduced the space but has lowered the performance.

The following figure 6.5 explains the snowflake schema in a diagram. This figure resembles shape that of a snowflake. It is this reason that it is termed as snowflake schema.

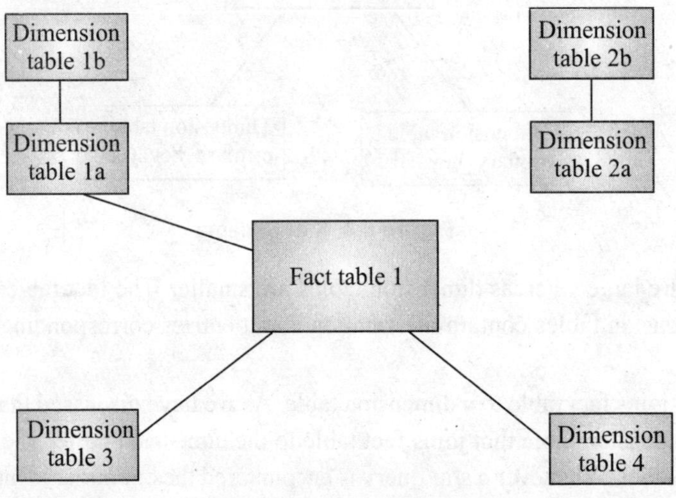

Figure 6.5. Snowflake Schema

In figure 6.5 we see the structure of a snowflake schema. In this figure we see that all the dimension tables are not joined directly to the fact table. Some dimension tables are joined to other dimension tables which in turn are joined to the fact table. So we can say that those dimension tables are joined to fact table through fact table. The number of dimension tables has increased. The advantage of snowflake schema is that the overall space requirement has decreased. The disadvantage of snowflake schema is that amount effort it requires has increased due to more number of look up tables.

6.5.5 Fact Constellation

Facts are the parameters that measure the business. A fact is a truth or something that has happened in past. For example the sale of a product measured in some figures is a fact. Or the turnover of a company in the last year is measured in amount and is a fact. Some data measured in numerics is dimension. However they seem to be fact. It would be efficient to classify the border line fields to be dimension.

We have already seen in the previous sections that fact table contains a number of primary key attributes from dimension tables. The fact table also contains many facts that are measurable and have their importance to business analysts, managers and executives. Let us take an example. The fact table of an entity employee will contain many entries. These entries are entered in rows. There will be as many rows as there are employees. The fact table for customer entity of a credit card company may contain millions of rows. Usually fact tables are designed to store numeric data which is measurable. Examples are salary, turnover, number of customers, etc.

There are many columns in a fact table that come under two categories. First category is the one that represents the facts in numerical form. The other category is the one that contains foreign attributes which are included as primary key attributes in the dimension tables. Some fact tables that contain merged or aggregated facts are also called summarized tables. This is because in these tables the facts are aggregated which give summarized results. Facts can be additive, semi-additive or non additive. Additive facts are those that can be added for example profit. Semi-additive facts are those that can be added partially along only some dimensions. Non additive facts are those that cannot be added for example percentage.

Now let us examine what is fact constellation. In the early classes we have studied in physics about constellation. Constellation is a collection of some stars that depicts a shape. Similarly, fact constellation is a collection of some fact tables. Fact constellation schema is the mostly used schema because it can also include interrelated and multiple subjects. The figure below shows a fact constellation.

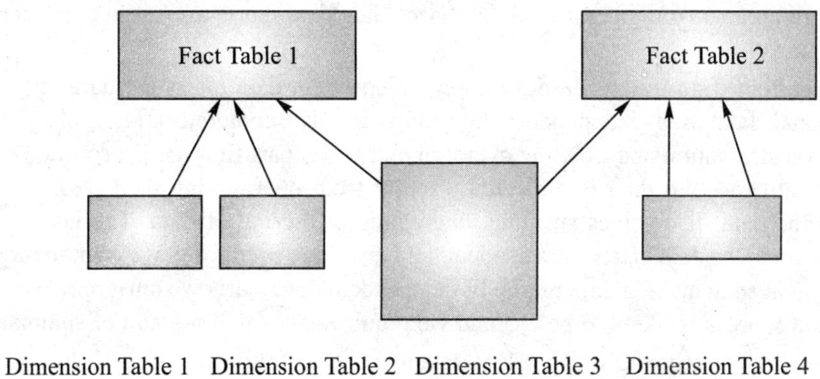

Figure 6.6. Fact Constellation

6.6 DATA MARTING

In this chapter we have studied about data warehouses. We have studied why we need data warehouses and why do we require separate warehouse for data analysis when we are given the operational databases. We are now going to study the data marts which are an important part in the discussion of data warehousing. We have already seen that warehouses contain the summarized data but still we need to divide the warehouses to smaller units known as data warehouses. This divides the problem and makes it easier to solve. Each data mart is applied to solve a specific problem. The data in data mart can be organized according to star schema. The facts are stored in a fact table at the center. The fact table remains joined to dimension tables on some foreign key attributes. Data in the dimension tables are the dimensions which act as parameters along which data analysis can be performed.

The benefit of data mart is that it is concentrated on problem of only particular subject. The simple data marts may get their data directly from databases without the involvement of warehouse. We have already seen that data mart is concentrated only on data of particular subject rather than complete data from database. If the strategy to make a data mart is adopted carefully then it helps in reducing the risk, cost and time. Data mart is not affected by size of organization or the problem complexity. If the problem is complex it is advisable to divide the problem into further smaller data marts which are responsible for small set of problem.

We have already seen that data marts reduce the risk associated with developing a system that gives capabilities to users for analyzing data online. Additional advantage is that it is very easier to implement. It does not require any detailed knowledge or skills to implement. The only principle that works is that if the problem is complex then disintegrates the problem further. The technology of data mart is new at present and there are only few experienced professionals that can effectively and efficiently use data marting. However because it is easy to learn and implement, the new users can effectively develop their skills by learning from experienced people.

There are two types of data marts: Dependent data marts and Independent data marts.

Dependent data marts: In dependent data marts the data for the data warehouse concerned is directly loaded from the warehouse to the data mart. As the data is already summarized in a data warehouse the effort required to develop a data warehouse is correspondingly reduced. For dependent data marts only the correct data source related to the subject is to be loaded in the summarized form. Dependent data marts are designed to reduce the costs and improve the overall efficiency and performance. One disadvantage of using dependent data marts is that it takes more overall time. This is because first data warehouses are created from operational databases then data marts are created by identifying related data warehouse.

Independent data marts: In developing independent data marts the data is extracted directly from operational databases. Independent data marts involve complete ETL cycle. ETL cycle is the Extraction, Transformation and Loading cycle. In extraction part first the appropriate data sources are identified. As already told data is collected directly from operational databases. The next phase is transforming the data. It involves splitting, aggregation, filtering of data. The last part is loading of transformed data in the data marts. Independent data marts are used when we want to reduce the number of overall time taken in making data marts. In independent data marts we must remember one thing that number of data sources is likely to be 2-3 and we require only small amount of summarized data.

Problem associated with independent data marts is as follows:

The creation of too much data marts creates redundancy problem. This is because each data mart requires a copy of data from database each time. Hence the number of redundancies tends to grow with the number of independent data marts. The other problem is associated with cost. When building independent data marts no central data warehouse is build. Hence each time a data mart is required complete extraction, transformation and load cycle is repeated again and again. This increases the duplicity and increases the redundancies. For doing this expert staff is required. Large numbers of people are involved who work on their own and use different tools which cause lack of generalization of techniques. This increases the time to normalize hence increasing cost and time.

6.7 DATA WAREHOUSE ARCHITECTURE

There are five types of architectures of data warehouses:

1. Independent data marts
2. Data mart bus architecture with linked dimensional data marts

3. Hub and spoke architecture
4. Centralized data warehouse
5. Federated architecture

1. **Independent data marts:** All companies do not use same data marts. They develop their data marts in isolation from other data marts. Hence these data marts are not in generalized format. They use different measures and dimensions giving rise too much inconsistency and makes it difficult for business professionals to analyze data.

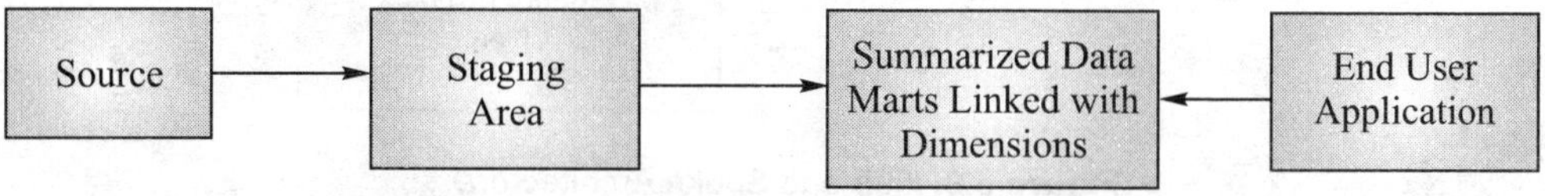

Figure 6.7. Independent Data Mart Architecture

2. **Data mart bus architecture with linked dimensional data marts:** The basis for data mart bus architecture with linked dimensional data marts is the complete study of all the business requirements in business process. For example in a library system the requirements study would emphasize on book issue, total fine, purchase of new books. In accordance with a business process the first data mart is created considering related dimensions and measures. Then other data marts are created in accordance with conformed dimensions and measures. The data marts give rise to summarized data which are arranged in star schema with fact table at the center and related dimensions.

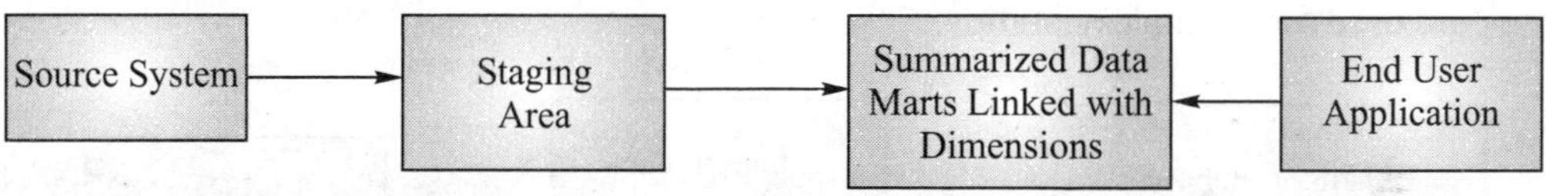

Figure 6.8. Data mart bus architecture with linked dimensional data marts

3. **Hub and Spoke Architecture:** The foundation for this architecture is the proper analysis of business requirements. This architecture is developed iteratively. First we concentrate on one subject. When this subject is completed then we consider the next subject. The data in warehouse is maintained in summarized form. This warehouse is used to take data for dependent data marts. The data marts created may have summarized normalized, denormalized data according to requirements. Figure 6.9 explains Hub and Spoke Architecture.

4. **Centralized Architecture:** The centralized architecture is almost similar to Hub and Spoke Architecture in all parameters. The only difference is that there are no dependent data marts. The warehouse contains data in atomic form and contains logical dimensional views of data. Both dimensional views and relational data are used for user applications and queries. This architecture is the logical implementation of the previous architecture. The diagram in figure 6.10 depicts centralized architecture.

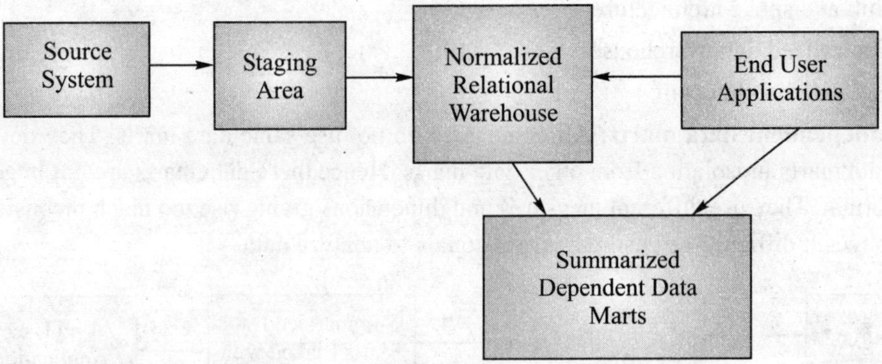

Figure 6.9. Hub and Spoke Architecture

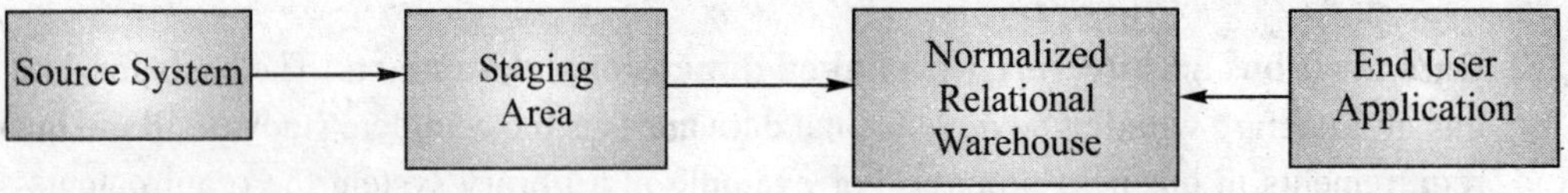

Figure 6.10. Centralized Architecture

5. **Federated Architecture:** Federated architecture is designed for complex business requirements. It is used in those cases for the companies that already exist. This architecture uses operational databases, data warehouses and data marts as data sources for developing an architecture which supports analytical processing. Meta data, shared keys and global queries are used for its implementation.

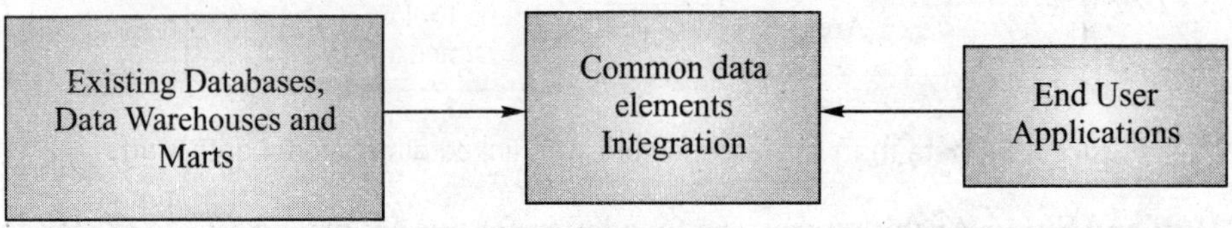

Figure 6.11. Federated Data Warehouse Architecture

6.7.1 Factors Affecting Data Warehouse Architecture: Mapping Warehouse to an Architecture

Different companies use different architectures. There are disagreements over which architecture should be used and which architecture should not be used. There are eleven factors in total which decide which architecture to be used. These factors are described as follows:

1. **Informational Interdependence between Organizational Units:** If one organization is dependent on information from some other organization then architecture that maintains information in the most consistent and integrated form is required. Hence the organizations that require high level of information interdependence require wide architecture.
2. **Upper Management's Information Needs:** In an organization the information from lower level employees is passed to employees at high positions. When this information is passed by lower level employees to higher level employees about the company aspects like turnover,

sales then lower level people collects information about some specific subject. The information may be filtered and aggregated. This facility also requires an architecture that supports it.

3. **Urgency of Need of Data Warehouse:** Sometimes an organization may require data warehouse urgently. So the first requirement is that time required to make the data warehouse should be less. Some warehouses can be implemented faster whereas some architecture requires time. Hence urgency of need of data warehouse is an important factor.
4. **Nature of End User Tasks:** The work of some employees is not same throughout the year. They have to manage different types of data, analyze the data. So these people require large amount of data that can be managed efficiently. So this facility requires the architecture that supports it.
7. **Constraints on resources:** Some architecture require more resources while some architectures require less resources. Hence resources like technical people, business people and money plays a key role in selection of architecture.
6. **View of the Data Warehouse Prior to Implementation:** Different organizations differ in their strategies for data warehouse. Some make it suitable to support some business need while some make it to support large number of applications. Hence on the basis of this different architectures can be used.
7. **Expert Influence:** Presence of expert professionals (experts, theoretical and practical users) also affect the selection of a particular architecture. IT consultant would want to use the same architecture that he has implemented in past.
8. **Compatibility with existing system:** The architecture that is required to be implemented should be compatible with already present computing systems. The architecture should be compatible with the technology, systems and user applications already in use.
9. **The Perceived Ability of the In-House IT staff:** Different people in accompany have different levels of technical skills, knowledge and experience. Hence these people will consider the problem with different level of complexity. Hence they will try to implement the problem with different types of architectures depending on their skills and encounter with the similar problem in the past.
10. **Source of Sponsorship:** When a data warehouse is implemented it may be sponsored by a department or top level management. Hence the architecture is affected by the source of sponsorship. Different sponsors will want to select a particular architecture on the basis of resources required or the requirements.
11. **Technical Issues:** The technical issues are also responsible for the selection of a particular architecture. Parameters like number of customers, scalability, metadata and performance are responsible for selecting a particular architecture.

These are the eleven factors that are identified to be responsible for selecting a particular architecture. On or the multiple factors are responsible at a particular time (some factors may be absent at some situation for selecting an architecture. The below diagram depicts the total factors and the total architectures. It should be noted that more than one factor are applicable at a time but only one architecture of the five is selected.

Data Ware House Architecture Factors	Architecture Selected
Informational interdependence between Organizatrional Units	Independent data marts
Upper Management's Information Needs	Data mart bus architecture with linked dimensional data marts
Urgency of need of Data Warehouse	Hub and spoke architecture
Nature of End User Tasks	Centralized data warehouse
Constraints on resources	Federated architecture
View of the Dat Warehouse Prior to Implementation	
Expert Influence	
Compatibility with existing system	
The Perceived Ability of the In-House IT staff	
Source of Sponsorship	
Technical Issues	

Figure 6.12. Factors Affecting Data Warehouse Architecture and Warehouse Architectures

6.7.2 Parameters for Assessing Success of Architecture

A number of parameters have been described that explain if a particular architecture was successful or not. Some metrics are based on quality, efficiency and performance. And some metrics are based on if the architecture was implemented in the given time. And information was complete or not that is all the requirements have been achieved or not. These six metrics are used to measure success of architecture.

The following are the parameters of measuring success:

1. **Information Quality:** Information quality is defined by three things: Information accuracy, completeness and consistency. Information accuracy means that warehouse should be accurate. Information completeness means that all the information that is required to be added in a warehouse has been added. Information consistency means that data in the warehouse should not be inconsistent.
2. **System Quality:** System quality again consists of three things- system flexibility, system scalability and system integration. System flexibility means that warehouse should be flexible and should adjust against the change of requirements. System scalability means that warehouse should be scalable to contain as many entries as possible. System integration means that it should be possible to integrate data from multiple sources.
3. **Individual Impacts:** The warehouse should improve the analytical capabilities and decision making capabilities of user. It should be easier to use and should help users in analyzing data.
4. **Development Time:** The development time is also an important parameter in analyzing if a particular architecture is successful or not. If architecture is implemented after decided time it is unsuccessful.

5. **Organizational Impacts:** The warehouse should fulfill all the business requirements for which it was used. Not a single requirement should be left.
6. **Development Cost:** The development cost should be under the decided budget. If the expenditure of the warehouse increases above the decided budget it is held unsuccessful.

6.8 DATA WAREHOUSE ARCHITECTURE IMPLEMENTATION

In the previous section we studied data warehouse architectures. We studied the five architectures and factors affecting data warehouse. In this section we will study the implementation of warehouse architecture. Three approaches are used to implement the architecture. These approaches are top down, bottom up, or a combination of both. The choice of approach depends upon current architecture, available resources, implementation scope and speed, return of investments.

1. **Top Down Implementation:** It requires more planning and design work at the beginning of project. People from work groups, departments, decision concerning data sources, security, data structure, data standards, data model is required before the implementation begins. The company wide warehouse with access to data is also required. Figure 6.13 below shows its implementation.

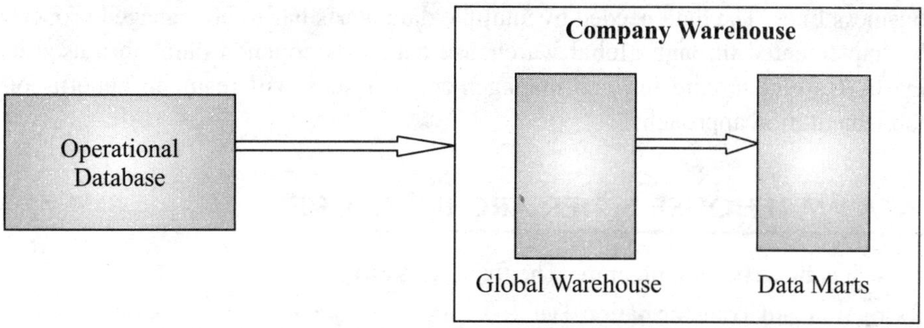

Figure 6.13. Top Down Implementation

This implementation results in more consistent data definitions at the expense of high cost of initial planning and design. Because this is a time consuming process hence may result in delay in actual implementation and benefits. Sometimes these delays may not be accepted. It works fine when hardware and computer resources are easily available.

2. **Bottom Up Implementation:** The bottom up implementation is different from previous implementation in the sense that it involves planning and designing of data marts before complete global infrastructure. The global infrastructure will be developed incrementally with data mart expansion. This implementation is better than Top Down in the sense that immediate results from data marts can be realized and used as justification for expanding global infrastructure. Data marts may be built in parallel with global warehouse. Data in data marts can be stored directly from operational databases or from warehouse. This implementation approach is faster and prevents delay. But one thing that needs to be taken care of is that anomalies and redundancies may occur which requires careful monitoring and guidelines. As the implementation grows some rework may be required to change existing implementation areas. These issues should be taken in care. The diagram in figure 6.14 represents bottom up implementation.

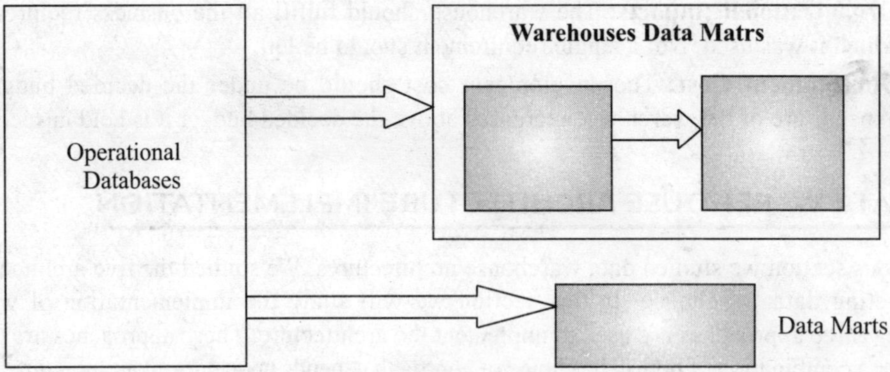

Figure 6.14. Bottom Up Implementation

3. **A combined approach:** We have seen both the approaches above. Both approaches have some advantages and some disadvantages. We would like to take benefit of the combination of the both approaches. The degree of planning and careful design is required because data marts are also built in parallel with the global infrastructure. For the global data warehouse develop the base level infrastructure definition. First step would be identify participating business lines. The data needed by multiple data marts has to be managed properly. It could be implemented through global warehouse acting as common data store used by all data marts. Careful monitoring and management of issues will result in benefits of both the implementation approaches.

6.9 DATA WAREHOUSE 3 TIER ARCHITECTURE

Data warehouse has three tier architecture. The three tiers are:

1. Extraction and Transformation Tier
2. Connective Tier
3. Data Access and Retrieval Tier

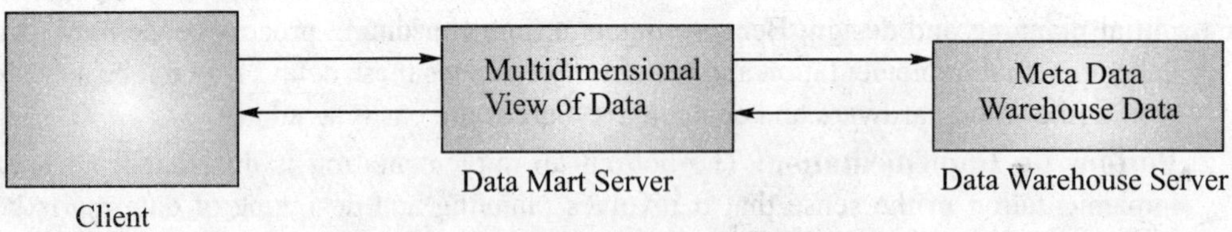

Figure 6.15. Data Warehouse 3 Tier Architecture

1. **Extraction and Transformation Tier:** Data is collected from various sources like company database, operational database and is then refined eliminate non useful data. The extracted data is then transformed into a standard format required according to problem. The transformed data is then loaded into the data warehouse. The process of extracting, transforming and loading is called ETL process. We have read about ETL in the previous sections.
2. **Connective Tier:** The data mart server serves as connective or middle tier. The data extracted and transformed from database is stored in warehouse which acts as central storage.

3. **Data Access and Retrieval Server:** The end user enters a query through OLAP tool. The query is processed is processed by warehouse and the graphical and complex results are displayed.

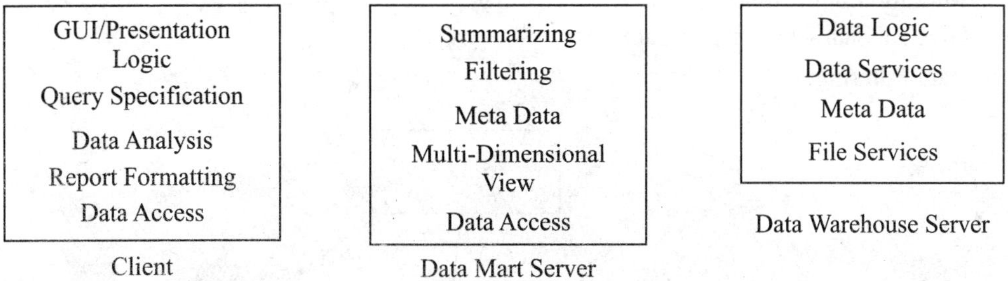

Figure 6.16. 3 tier Architecture

6.10 DATA WAREHOUSING TO DATA MINING

Both data mining and data warehousing are business intelligence tools that are used to turn information into actionable knowledge. However there are some differences between methods and processes of both that are used to achieve their objective. Data mining is a process of statistical analysis. Different tools are used to process huge amount of data. Analysts assume that if a customer buys a product then it is probable that he will buy some other but related product during a course of time. Running a query on a particular data to prove or disprove this assumption is data mining. This information is critical for business management to decide their strategies for future.

Data warehousing on the other hand as discussed in the last chapter is a technique in which data from various business and operational databases is stored in the data warehouse so that it is easier and convenient to analyze huge volumes of data. Data is aggregated in the data warehouses. Data from various stores is connected conceptually as well as physically in warehouses.

So the relationship between data mining and data warehousing is that if the data is warehoused then it is easier to mine this data. If data mining query were to run on different sources of data then data accessing and processing would take time resulting in inefficiency. But if a proper warehouse is built where data from various sources is connected in summarized form then it is easier to run a mining query.

6.11 METADATA

Metadata is an important element of data warehouse. Meta data is data about data. We are familiar with metadata concept used in advanced java. The role of metadata is very important. Its use in data warehouse is new.

The role of metadata in data warehouse is big because it is used in mapping data from data operational database to data warehouse. For example suppose we need to make a analytical report on some subject. The report can be made from the summarized data stored in data warehouse. But suppose if we are also required to validate from operational database if the report is correct. This can be checked only from mapping of operational database to data warehouse. Now where is this mapping stored. This mapping is stored in data warehouse store. The components stored in metadata store are tables of data warehouse, keys of these tables and attributes of tables.

Mapping includes:

1. One attribute to other attribute
2. Conversion
3. Alteration in naming
4. Physical characteristics of data
5. Data Filtering

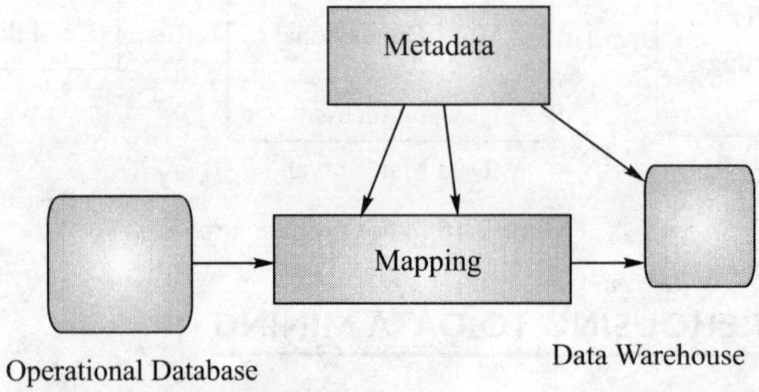

Figure 6.17. Mapping of Operational Database and Data Warehouse

6.12 CONCEPT HIERARCHY

Concept hierarchy is an order relation between set of attributes of concept or dimension. The concept hierarchy can be generated by users or automatically. We have studied about multi-dimensional data. Multi dimensional data is organized into dimensions. Concept hierarchy is used to provide lower level abstraction of a dimension. It is used in summarization of data cube. Concept hierarchy maps lower level concepts to more general higher level concepts.

Let us try to analyze the concept hierarchy by an example. Suppose a company has its offices in Hisar, Gurgaon and Pune. Now Hisar and Gurgaon can be mapped to higher and generalized concept Haryana and Maharashtra, the corresponding states. These sates could be mapped to more generalized India which in turn could be generalized to Asia. Similarly an office in Kuala Lumpur will be mapped to Malaysia which is mapped to Asia. This is explained diagrammatically in figure 6.18.

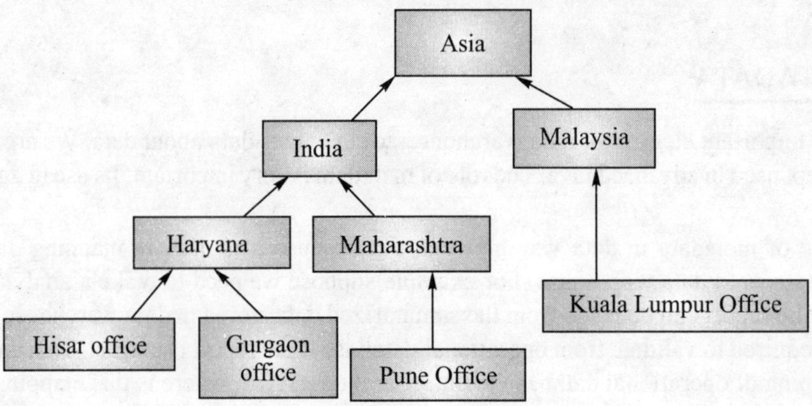

Figure 6.18. Example of Concept Hierarchy

SUMMARY

- A data warehouse is a subject-oriented, integrated, time-variant, non-volatile collection of data in support of management decisions.
- Data Warehouses are subject oriented, integrated, non volatile and time variant.
- Warehouses are required for analysis of historical data. It is also required to plan the future strategies.
- The warehouses include the entire delivery process for making a data warehouse.
- Data Warehouses are different from operational databases in the sense that operational databases store the entire data whereas warehouses only store the aggregated data.
- The components of data warehouses are identifying the source, cleaning up and transformation and migration tools.
- The multi dimensional model of warehouses consists of facts stored along the various dimensions.
- There are three schemas of data warehouses namely star schema, snowflake schema and fact constellation.
- There are two types of data marts, Independent Data Marts and Dependent Data Marts.
- There are five types of data warehouse architectures Independent, Data mart bus architecture with linked dimensional data marts, Hub and spoke architecture, Centralized data warehouse and Federated architecture.
- The selection of a particular architecture depends on various perspectives.
- There are three types of implementation strategies of data warehouse architecture top down, bottom up and mixed approach.
- There are three tiers of 3 tier data warehouse architecture.
- Both data mining and data warehousing are business intelligence tools that are used to turn information into actionable knowledge
- Metadata is data about data.
- Concept hierarchy is an order relation between set of attributes of concept or dimension.

REVIEW QUESTIONS

1. Define Data Warehouse. Explain the features of data warehouse.
2. Why do we require data warehouse?
3. What are the steps in the delivery process of data warehouse?
4. Differentiate between database and data warehouse.
5. What are the components of data warehouse?
6. Write short notes on:
   (*a*) Star schema
   (*b*) Snowflake schema
7. What are the factors affecting data warehouse architecture?
8. Explain the concept of concept hierarchy.

CHAPTER 7

Data Warehousing and OLAP

In the previous chapter we studied about some concepts of data warehousing. We studied why data warehousing is so important in today's software industry. We studied the features of data warehousing. We also studied various schemas and the data warehouse architecture, factors affecting the selection of an architecture and metrics to analyze success of the architecture. We also studied the data model in data warehousing and multi-dimensional data model. In this chapter we will study some other important concepts of data warehousing. We also look at some important concepts of OLAP.

In this chapter Section 7.1 describes the Aggregation. In Section 7.3 we will focus on data cubes. Section 7.4 describes the query facility in warehouses and how it is implemented. In Section 7.5 our focus would be to study OLAP that is online analytical processing. From Section 8.6 to Section 7.9 we will study various concepts involved with OLAP. It includes OLAP patterns and modals, OLAP functions, OLAP tools and OLAP servers.

Section 7.10 describes data mining interface. Section 7.11 and section 7.12 enlighten Data warehouse security and backup and recovery in warehouses. In Section 7.13 we will study how to tune a data warehouse. Section 7.14 describes important aspects of testing of data warehouses.

7.1 AGGREGATION

We have used data aggregation many times in the previous chapter. We are familiar with the meaning of aggregation. Aggregation is nothing but a technique to merge or consolidate dispersed data into single data. The data that has been summarized is the summarized data. Aggregation plays a key role in data warehousing. Aggregation is done to speed up the query processing and improve the overall performance in terms of query response time and data warehouse usability. Aggregation is used when the data extracted from databases has to be transformed to be stored in the data warehouse. Aggregation summarizes different data into a summarized data by removing duplicate data. One disadvantage of aggregation is that although the queries are run faster yet some data may be lost.

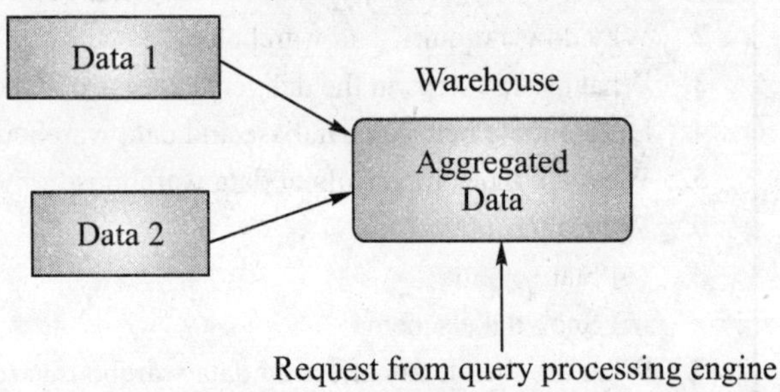

Figure 7.1. Aggregation in warehouses

7.2 HISTORICAL INFORMATION

In the previous chapter we have used historical information keyword many times. We have also studied this in the previous chapter in the beginning. The warehouses are based on this historical information. It is this historical information that is used to develop a mechanism to support decision making strategies.

Let us look it in this way. Operational databases store every operation or transaction. This information is stored in the database at the time the transaction is performed. Now when we are storing the data from databases to the warehouses this data acts as historical information which acts as a base for performing decision making. This historical information is however stored in warehouses in the summarized or aggregated form. Hence the historical data is not any data related to ancient India. Here it is related to the data stored in operational or transactional databases.

7.3 DATA CUBES

We have used the term data cubes in the previous chapter. Some of you might have understood the basic thing about data cubes. In this section we will study data cubes in detail. How they are used in data warehousing and what is their importance in OLAP. We will study OLAP later in this chapter. For we just know about the basic overview of OLAP. OLAP is online analytical processing. OLAP is used to process hundreds of queries over large volumes of data. OLAP is used for decision support functions. For now this basic idea is sufficient to understand data cubes on which we are going to focus in this section.

Data cubes represent multi dimensional data. Multi dimensional data is represented by facts and dimensions. Dimensions are the parameters or perspectives along which data is stored. Data cubes become important when we store multi dimensional data. Multi dimensional data means data along many dimensions or perspectives. Multi dimensional model focuses on entities like loan by a bank which is stored in a fact table. Facts are stored numerically in a fact table. A fact table may contain entries from few hundreds to millions of entries. These facts are recorded along some dimensions. Now whenever a user asks some question then this question is converted to a query by query processing engine. Now there could be hundreds of users giving rise to large number of queries at the same time which are difficult to processing by scanning the data every time. Hence we require some method that provides quick access and processing. Data cube is the answer to our problem. Data stored in data warehouses is quickly accessed through structures called data cubes. The choice of structure of data cube is made by analyzing user requirements and the choice of schema. We studied three schemas in the previous chapter namely star schema, snowflake schema and fact constellation. The Fig. 7.2 represents sale in lacs by several companies in different countries over the years.

There are three companies X, Y and Z arranged along the x-axis. There are four countries India(I), Pakistan(P) and Bangladesh(B) shown in figure which are arranged along the y-axis. The two years 2010 and 2011 are shown along the z-axis. The intersection of an each element from x, y and z axis gives the sale in lacs of a particular company in a particular country in a particular year. 'All' given along an axis displays the sum of sale in all entities with the intersecting dimension.

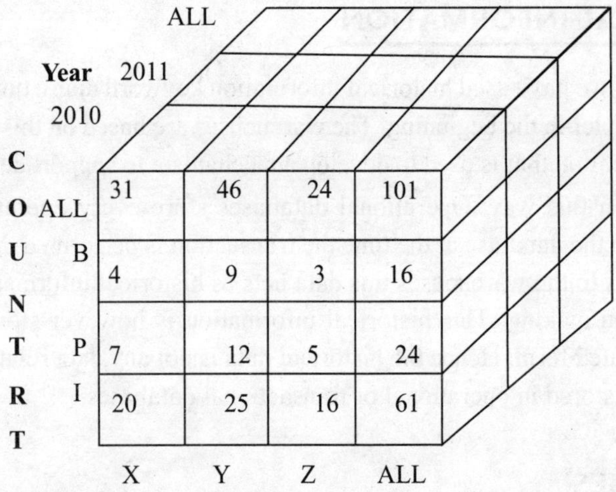

Figure7. 2. Data Cube

In the three dimensional cube given above three types of queries or aggregations are possible:

1. null (*e.g.*, what is the total sale? Only 1 possible query)
2. companies (*e.g.*, what is the sale by company X? 3 queries are possible as there are 3 companies)
3. country (*e.g.*, what is the sale in India(I)? 3 queries are possible as there are 3 countries)
4. year (*e.g.*, what is the sale in 2010? 2 queries are possible as there are 2 years shown)
5. company, country (*e.g.*, what is the sale by company X in India? 9 queries are possible, 3 companies and 3 countries)
6. company, year (*e.g.*, what is the sale by Y in 2011? 6 queries are possible, 3 companies and 2 years)
7. country, year (*e.g.*, what is the sale in Bangladesh in 2010? 6 queries are possible, 3 countries and 2 years)
8. company, year, country (*e.g.*, what is the sale by X in 2010 in Pakistan? 18 queries are possible, 3 countries, 3 companies and 2 years)

Hence data cubes provide an easy way to access various levels of data present in the database in a faster way.

7.3.1 Data Cube Operations

The common operations that could be applied to data cubes are:

1. Roll Up
2. Drill Down
3. Slice and Dice
4. Pivot

1. **Roll Up:** The Roll Up in data cubes uses concept hierarchy. We have studied concept hierarchy in the previous section. We have seen how concept hierarchy maps lower level details to higher generalized details. We have seen an example where an office in a city is mapped to the state which in turn is mapped to higher details. As we summarize the data the cardinalities shrink resulting in smaller cube. Hierarchy defined on a dimension is useful in the roll up operation.

2. **Drill Down:** Drill Down is just the opposite of roll up. Drill Down goes from higher level details to lower level details. Here we can use concept hierarchy but we move down the hierarchy. Drill Down is just like zooming on a point. For example we get a point or a topic and we search down every single detail about it.
3. **Slice and Dice:** This is used for searching and accessing data in the cube. Let us try to explain the meaning the meaning of Slice through an example. Suppose we have a three dimensional equation. A three dimensional equation uses three dimensions. But now if we assign a value to one of the variables then we get a two dimensional equation. In data cubes this two dimensional equation could be mapped to the data after slice operation.
4. **Pivot or Rotate:** This operation is used when a user wishes to change the orientation of the view of the cube. In this operation position of some rows or columns may be changed.

7.4 QUERY FACILITY

Many types of simple business questions can be expressed as SQL queries. For example number of students studying B.Tech in India in a particular year. SQL is highly used as a language that supports queries. SQL in warehouses is implemented similar to it is implemented in databases. In data warehouses it consists of OLAP functions that can be used to solve complex queries that need to calculate ranks of the people on the basis of some attribute. We will study OLAP functions later in this chapter where we will see various functions and their benefit.

However comparison of various data values cannot be expressed as a query by user, query language and database server. Questions that involve sequential processing are difficult to write as SQL queries.

To solve this difficulty involved several OLAP tools are built that provides excellent interface and query processing scheme. DMQL Data Mining Query Language is used to write queries in data mining and data warehousing. Similar to RDBMS there exists DDL to define tables in warehouses.

Cube Definition

Syntax:

```
define cube <cube_name> [<dimension_list>]:
    <measure_list>
e.g.
    define cube loan [customer, employee, branch]:
    total_loan = sum(loan in dollars),
    total_loan_customers = sum(loan customers)
```

Dimension Definition

Syntax:

```
define dimension <dimension_name> as (<attribute_list>)
e.g.
define dimension customer as (cust_id, cust_name, loan_amount, address)
```

Different type of queries can be implemented and defined by DMQL. On the basis of above example all the schema can be implemented in DMQL.

7.5 OLAP-ONLINE ANALYTICAL PROCESSING

We are now about to discuss the OLAP. We have used OLAP at some places in the previous chapter and this chapter. By now you must have gained some insight about OLAP. OLAP is an important concept in data warehousing. OLAP tools provide effective interface to analyze multi dimensional data. Association, clustering, classification and prediction along with OLAP provides mining of knowledge at multiple levels of interaction. The reason behind growing of OLAP is the growing business and e – commerce. The analysts analyze the summary of the transactions taking place in the operational databases. We are going to study OLAP in detail in this chapter.

7.5.1 Introduction

OLAP allows the business users and executives to access data. An organization may have many centers all over the world. The customers buy online products all over the world. Now the information about these transactions is stored on servers of these centers. Now if the CEO of the organization wishes to see what is the sale of a particular product then it would be inefficient to analyze the data from all these centers individually then add them.

The data from all the centers all over world is stored in the warehouses in the aggregated or summarized. This stored data is in multi dimensional form as discussed in the previous chapter. OLAP provides an effective way to analyze the data. The question like what is the sale of a particular product is converted into query and the result is displayed in front of the user. Hence it is very easy to process large amount of data. Hence there is no need of processing query on individual data stored at each server. OLAP is the dynamic enterprise analysis required to create, manipulate, animate and synthesize information from data models.

OLAP implementation process involves extracting data from various data sources. It means that data at one source matches data in all the sources.

It should be noted that data warehouses may not be created always for analysis by OLAP. There are the databases called OLTP called online transaction process databases. These are the databases where the information regarding customer_no, order_no, date, product_id, payment option is stored. This stored is stored immediately when a customer makes an online purchase. This data is accessible to analysts as soon as the transaction is over.

OLAP is different from OLTP in the way that OLAP gives a summarized view of what happened. The most common data storage of OLAP is star schema which we studied in the previous chapter.

The characteristics of OLAP are as follows:

1. **Multidimensional Conceptual View:** We have already discussed what the multidimensional data means. It helps in carrying slice and dice operations.
2. **Accessibility:** OLAP acts as a mediator between data warehouses (back end) and front end.
3. **Multi User Support:** The OLAP system is shared. Hence OLAP software must be compatible for retrieval, update, concurrency control, back up and security.
4. **Storing OLAP Results:** OLAP results are kept separate from data sources. OLAP applications should not be applied on live transactional databases.
5. **Extraction of Missing Values:** OLAP provides for distinguishing between zero values and missing values so that aggregates are computed correctly.
6. **Treatment of Missing Values:** OLAP system should ignore all missing values and compute correct aggregate values.

7. **Uniform Reporting Performance:** If the size of the database is increased then it should not hamper on OLAP performance.
8. **Generic Dimensionality:** All dimensions are equal in structural and operational capabilities.
9. **Unlimited Dimensions and Aggregate Levels:** There should be no limit on dimensions and aggregate levels.

7.6 PATTERNS AND MODELS

Model assesses the current situation (whatever the application is) and takes action. Models help in making decision making processes. Let us try to understand what patterns and models mean by an example. Suppose we are given a sequence ababa.... . Now suppose we are to find the next element of the sequence. The next element would be clearly b. But finding the next element in the above sequence is easy because the pattern is easy. The predictive model in the above example uses the algorithm that if the last element is a then the next element is b otherwise next element a. If the pattern is complex then it would not be that easy to evaluate the next element of the pattern.

The above example shows how a model can be used by analyzing the historic data and thus making a calculated guess of what the next sequence would be assuming that patterns are repeated. This helps in making decision making strategies. We saw an example in the last chapter of travel agency that plans its current year strategy on the basis of last year data of the number of people going to Shimla. In this way whatever the situation is it would be possible to monitor the historic data and plan the future strategies. Let us analyze the situation with another example. Suppose there is a worldwide shopping website that operates in selling products online. The business executives of the website would analyze the sale of their products in the past and hence would plan which products would be sold for the next month and at what prices. A model can be anything from mathematical equation to the set of rules that describe customer segments to complex neural network architecture. The figure below shows how a model is linked to database and prediction.

Let us now look into the formal definition of models and patterns:

Model: A description of the original historical database from which it was built that can be successfully applied to new data in order to make predictions about missing values or to make statements about expected values.

Pattern: A sequence of events that occurs in the database often.

The example that we studied previously in this section follows from the above given definitions. A pattern is something in which some event or some combination of event occurs again and again so that it is possible to recognize a missing event. Model analyzes the pattern and predicts the possible event. Patterns like some sequence of numbers are mapped to the graphical representation which helps humans to analyze the patterns easily. Complex sequence of numbers are easy to be analyzed by graphs. In a graph the value the y-axis at some point depends on value of x-axis or some previous value of y-axis. However in practical situation the value at some point may depend upon a number of factors.

In the Figure 7.3 the database corresponds to the data on the basis of which prediction is to be made. Record refers to the row in the table of a denormalized database. Prediction refers to the predicted data.

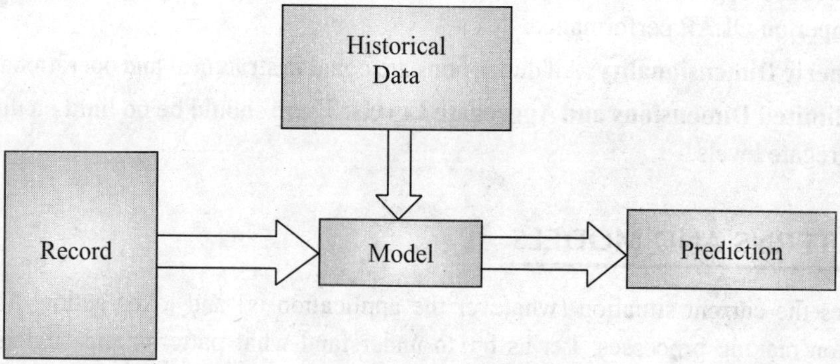

Figure 7.3. High level View of the Modelling Process

7.7 OLAP TOOLS

Today there a number of OLAP tools from a number of vendors are available in the market. The features and compatibility differ in some perspectives. The table below shows some of the common available OLAP tools. Some of these tools are supported by ROLAP, some by MOLAP and some by HOLP. We will study later in these chapter what ROLAP, MOLAP and HOLAP mean. For now it is sufficient to know that they are OLAP servers.

OLAP Server	*Company*	*Version*	*Software License*
Essbase	Oracle	11.1.2.1	Proprietary
icCube	MIS Consulting SA	2.5	Proprietary
Microsoft Analysis Services	Microsoft	2012	Proprietary
Micro Strategy Intelligence Server	Micro Strategy	9	Proprietary
Mondrian OLAP Server	Pentaho	3.4.0	EPL
Oracle Database OLAP option	Oracle	iigR2	Proprietary
Palo	Jed ox	3.2 SR3	GPL v2 or EULA
SAS OLAP Server	SAS Institute	9.3	Proprietary
TM1	IBM	10.1	Proprietary

Figure 7.4. Some Common OLAP Tools

7.8 OLAP FUNCTION

In this section we are going to study about OLAP function. OLAP functions have the capability of returning rank, row number and column function information as a scalar value in a query result. OLAP function can be included in the select list or the ORDER BY clause of the select statement. OLAP function is applied to the result table of the inner most sub select that includes the OLAP function. The OLAP function is specified with a window defining the rows over which OLAP function is applicable with their order. OLAP function can also be used with column. A number of preceding and following rows can also be specified. For example, average of the salary may be computed over previous two years salary along with the current year salary.

The ranking function is used to compute ranking of a row in the window. The same rank is assigned to rows that are not distinct with respect to ordering in the window. The rank of a row is computed by adding one to the number of rows that preceded the current row.

DENSE_RANK computes the rank of a row by adding 1 to the number of distinct rows that preceded the current row.

ROW_NUMBER computes the row number of the row in the window ordering by assigning 1 to the first row and plus 1 the subsequent row. But in case ORDER BY clause is not used in the window then row numbers are assigned in arbitrary order as returned by subselect.

The data type of all the three functions RANK, DENSE_RANK and ROW_NUMBER is BIGINT.

PARTITION BY is a function used to partition the result set.

ORDER BY *(sort key expression)* is used for ordering of rows in a partition.

(Sort key expression) defines the ordering of rows in a window partition.

ASC is used to ascend the values of sort key expression and DESC is used to descend the values of sort key expression.

There are several other OLAP functions also. But we don't need to study those functions right now. Let us try to understand the concepts by an example below.

Display ranks of employees in order by first name, according to the salary for employees earning more than Rs. 20000.

The solution query for this problem would use the columns that are required to be displayed, calculate ranks of employees on salary by RANK function by applying the condition on employees that salary should be greater than 20000 and then display the result in order of first name.

SELECT E_NO, FNAME, LNAME, SALARY, RANK() OVER (ORDER BY SALARY DESC) AS RANK_SAL FROM EMPLOYEE WHERE SALARY > 20000 ORDER BY FNAME.

Let us analyze the individual parts of the above query.

SELECT E_NO, FNAME, LNAME, SALARY, RANK(): It selects the columns values that are required to be displayed corresponding to the selected tupples.

RANK() OVER (ORDER BY SALARY DESC): It is a function that calculates the rank. ORDER BY SALARY DESC arranges the tupples in descending order of salary. RANK() calculates the RANK of these tupples.

AS RANK_SAL: RANK_SAL is the name given to the statement RANK() OVER (ORDER BY SALARY DESC).

EMPLOYEE is the table name.

WHERE SALARY > 20000: It is the condition that separates all those tupples that have salary greater than 20000.

ORDER BY FNAME: This clause is used to arrange all the selected tupples in the order of their first name.

In this way all the queries are possible to be written in data warehousing too. OLAP functions provide an effective way of writing complex queries so easily. It supports a number of features like calculating ranks.

Suppose the problem is as follows:

Display ranks of employees in order by ranks, according to the salary for employees earning more than Rs. 20000.

SELECT E_NO, FNAME, LNAME, SALARY, RANK() OVER (ORDER BY SALARY DESC) AS RANK_SAL FROM EMPLOYEE WHERE SALARY > 20000 ORDER BY RANK_SAL.

7.9 OLAP SERVERS

The chief component of OLAP is the OLAP server. OLAP server sits between a client and database management system. OLAP server understands the organization of data in the database and has special functions for analyzing the data. Today there are a number of OLAP servers available for almost every database management systems.

Figure 7.5. OLAP Servers

The selection of an OLAP server depends upon a number of factors. Different questions that arise when selecting an OLAP server are like what would be the cost of OLAP server, is OLAP server actually required, what are the memory requirements, etc. The figure 7.6 shows such questions.

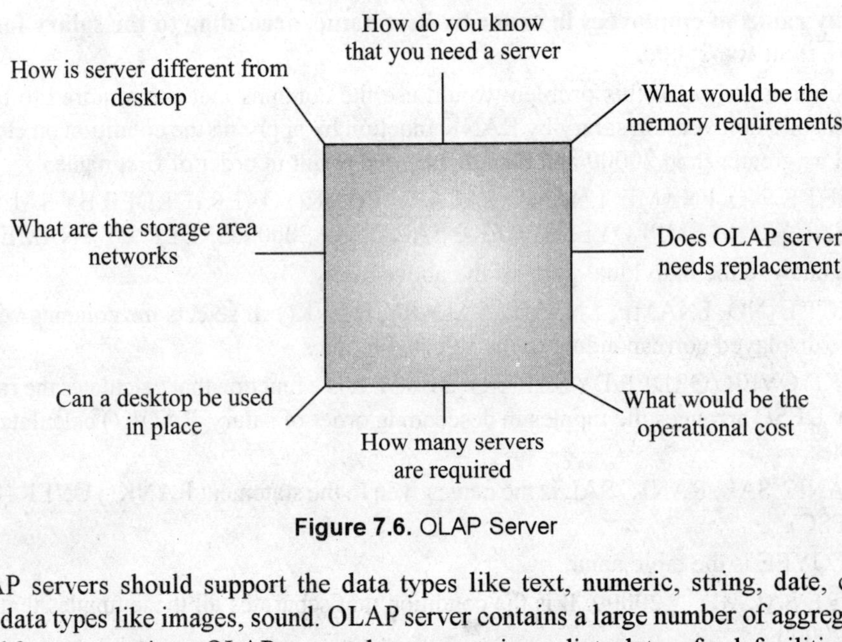

Figure 7.6. OLAP Server

OLAP servers should support the data types like text, numeric, string, date, calendar and multimedia data types like images, sound. OLAP server contains a large number of aggregate functions that are used for computations. OLAP server also supports immediate data refresh facilities for updating data load. OLAP supports multi dimensional data. The SQL extensions of OLAP operations are also implemented. Some of the supported SQL extensions are as follows:

1. **Extending the family of aggregate functions:** All the primitive functions used to in SQL like count(), sum(), avg(), min(), max() are supported. OLAP requires some other functions like RANK(), DENSE_RANK(), ROW_NUMBER(), etc. We have seen these functions in the previous section with example where we wrote a query to calculate ranks of employees on the basis of their salary.
2. **Adding reporting features:** OLAP system should support all the facilities like totals, cumulative totals and calculating averages for decision support properties.
3. **Implementing multiple group-by:** Data warehouses contain the multi dimensional data. So OLAP should support multiple group by on different attributes.

There are three types of OLAP servers:

1. ROLAP
2. MOLAP
3. HOLAP

1. ROLAP: ROLAP stands for Relational Online Analytical Processing. It uses relational database management system for implementing OLAP. ROLAP uses a bottom-up approach and uses a star schema data warehouse. We have studied the star schema of the data warehouse in the previous chapter. Hence the data in the database is in the denormalized state. A denormalized database contains many anomalies and redundancies but it grants speed and performance. We will again see the star schema of the data warehouse containing a fact table connected to multiple dimensional tables. Hence data warehouse supports multi dimensional data. The OLAP tool is used to manipulate the data of these tables. It calculates the aggregates and uses these aggregates to calculate new aggregates. The primary advantage of ROLAP tools is that it supports relational DBMS. The disadvantage is that the performance of query is low.

Below here is the star schema. The two types of tables are fact tables and dimension tables:

Fact table is used to store facts. The number of entries can be few to a million. The fact table contains primary key attributes from dimension tables that acts as reference key in fact table.

Dimension tables are used for storing dimensions along which the facts are stored. Each dimension table contains a primary key attribute.

The figure 7.7 depicts the star schema:

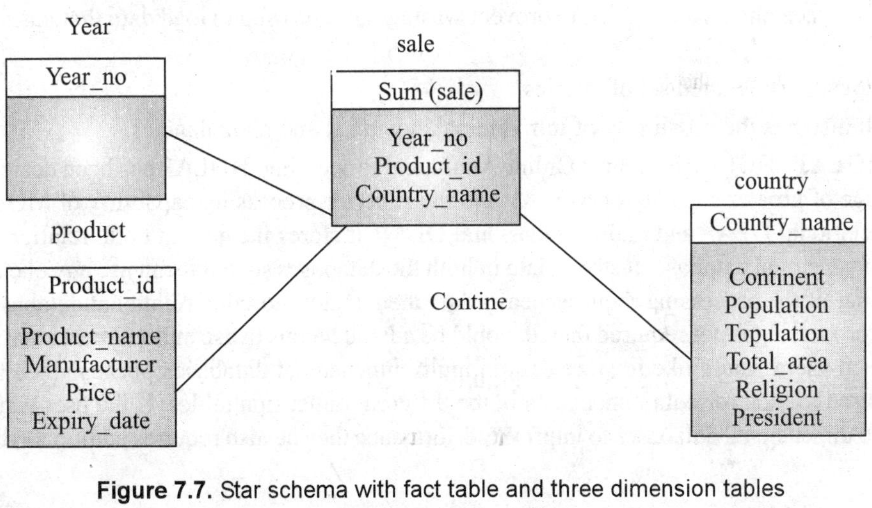

Figure 7.7. Star schema with fact table and three dimension tables

To normalize the star schema, snowflake schema was introduced. In snowflake schema, some dimension tables may not be connected directly to the fact table but through some dimension tables.

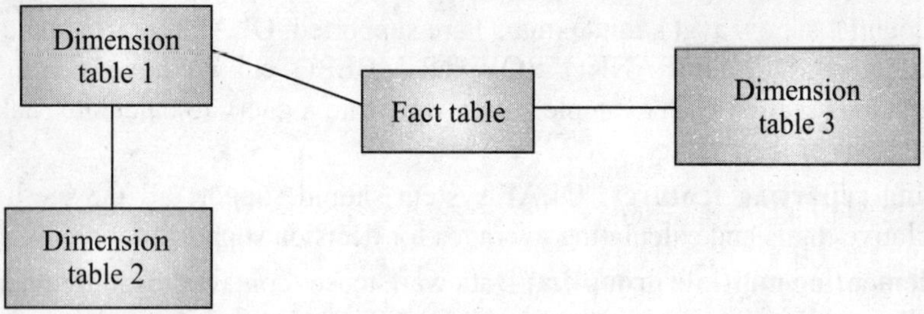

Figure 7.8. Snowflake schema of data warehouse

2. MOLAP: MOLAP is multi dimensional online analytical processing. MOLAP uses multi dimensional database to access data. Whereas the ROLAP uses bottom up approach, MOLAP uses top down approach. Unlike ROLAP, it uses special purpose file systems. We have studied the multi dimensional cubes where the data is stored along various dimensions. We saw an example where sale of a particular product in a country in a particular year was shown in a three dimensional cube. It requires large space to store multi dimensional data in cubes. Data retrieval and accessibility is fast. It supports limited set of queries. In MOLAP data can be stored in a relational database multi dimensionally but data is not accessible in a parallel way but individually from different tables.

Because the MOLAP processes data in a multi dimensional array, this results in faster data access and high performance.

Advantages of MOLAP are as follows:

1. It supports high query performance because of multi dimensional indexing.
2. The data stored in relational database takes much more space whereas here the space requirement is not much.
3. Computes higher level aggregates from lower level aggregates automatically.
4. Data extraction and accessibility is easy and fast.

Disadvantages of MOLAP are as follows:

1. The data load is slow. Hence it takes too much time to load update data. So only the data that has been changed is loaded to prevent wastage of time used to load data that has not changed at all.
2. It supports limited set of queries.
3. It also has the possibility of introducing anomalies and redundancies.

3. HOLAP: HOLAP is Hybrid Online Analytical Processing. HOLAP has been designed to take the advantage of greater capacity of ROLAP and the superior processing capability of MOLAP. It is a hybrid of relational OLAP and multi dimensional OLAP. It stores the data in both, relational database and multi dimensional database. It stores data in both the databases so that it can use any one as required for processing. If the processing requires heavy data then data is stored in relational database but if the heavy data processing is not required then it would be advantageous to use multi dimensional databases. The untrained users would like to store data in multi dimensional databases because then they would not be required to look for data dimensions in the different dimension tables. If the user wants to store data in multi dimensional databases to improve performance then he also requires additional data storage facility to accommodate multi dimensional data. But if the storage space is less then he will have to switch to the relational databases.

But still because HOLAP provides capabilities to support both MOLAP and ROLAP it is excessively used. If the user requires only aggregated data, a data cube of HOLAP may be preferred but if the detailed data is required by user then relational database may be used.

HOLAP cubes are smaller than MOLAP cubes and faster than ROLAP cubes. To support both ROLAP and MOLAP features their requirements must also be fulfilled.

MOLAP requirements are:

1. Faster access of aggregated data.
2. Easy maintenance of aggregated data.
3. Aggregated data should take less space.

ROLAP requirements are:

1. Refreshing of data dimensions on updation.
2. Should point to appropriate relational tables and should support automatic sql queries.

The data to be used for aggregation is stored in relational tables whereas aggregated data is stored in multi dimensional cubes.

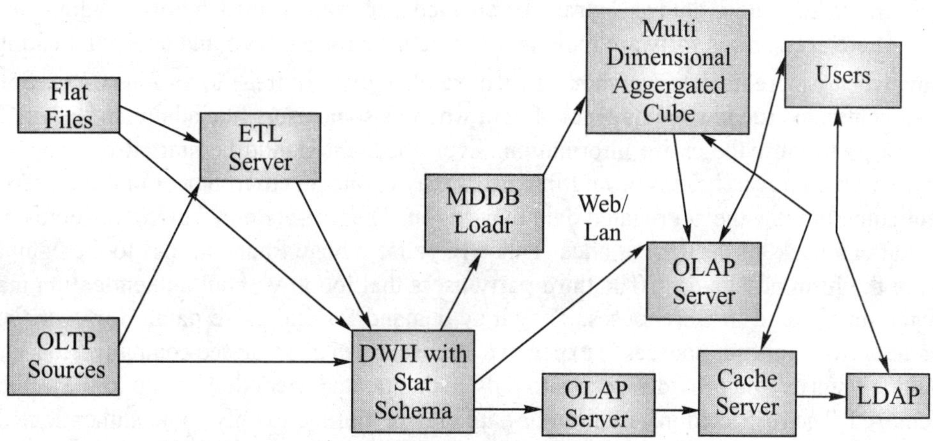

Figure 7.9. HOLAP Architecture

7.10 DATA MINING INTERFACE

Data mining is an exploratory process. An easy user interface is essential so that even non technical user can also use the data mining tool effectively. The graphic user interface should be of high quality and should be highly interactive because otherwise it would be much difficult for business users to understand it properly. Hence the GUI should be user friendly. Other user interfaces are constructed on top of SQL that serves as query language. But data mining interface does not share any data mining query language. Due to this data mining products are not standardized because there is no any common query language.

Let us generalize the functions of the data mining user interface.

1. The mining interface should be GUI.
2. There should be simple options for business or unskilled users but advanced options for skilled users.
3. Ability to build models in batch mode.
4. Additional applications should be supported by an API(Application Programming Interface).
5. The interface should also include help system.
6. The interface should provide default values to support lower limit records required to support a split in decision tree or number of nodes in hidden layer of NN so that probability for success is maximum for casual users.
7. Interface should provide management of data mining algorithms.

7.11 SECURITY

We have studied various aspects of data warehousing like data warehouses, schema, architecture and data cubes. We also studied how data in aggregated form as well as in detailed form is stored and how it is used to support decision making strategies. It helps business users to access data and make decisions.

Data warehouse as studied earlier acts as a data storage in aggregated form from various data sources. We have seen how data is extracted, cleaned and transformed before loading in the data warehouse. There are various software tools that can be used for analysis and decision making.

Now when we create data warehouses then we also give privilege to various users. Some users might be authenticated to view only a level of data whereas some users like administrator or CEO will be authenticated to view the entire information. Every account should be started by proper session. However the web data warehouses or multiple users may create an environment that the users that are not authenticated to view the aggregated data may log in. These users may belong to some competitor company that can hack all the data at once. This will cause a huge financial loss to the company. The other loss in the form of data cost. The third party users that log in without authentication may log in and view the data stored in warehouses. They may demand for excessive data. However the cost of extracting data from multiple sources is expensive and results in a loss to the company. Hence a system that maintains security and provides access to only authenticated users that too up to the authenticated level is required. The restricted information or data may be limited to only some authenticated users.

The following points must be kept in mind to maintain warehouse security.

1. **Replication Control:** First of all we should understand how data replication takes place. The data is replicated to the database or data warehouse if the particular data becomes outdated and hence the updated data is stored. One way would be to delete the old data and store the updated data. The other way that is used most commonly is to store both but store them as old version and new version of the data. Hence there are two versions of the data. Now if efficient management is not done on the old version of data then it would be easy for unauthenticated users to breach in and view the data. This data would be outdated but may prove very efficient for the business professionals in the rival company. Hence proper management of replicated data is required. Some users may keep on using the old version of data whereas some users that are willing to pay extra charges may be authenticated to view the new version of data.
2. **Aggregation and Generalization:** We have studied how data is stored in the data warehouses. The data in the warehouse is in the aggregated form. This data is summarized form of the raw

data present in the operational databases. This system of data may hamper the security because if the data is stored in the form of records then some unauthenticated users may breach the security and hack that data. Hence the system of generalization of data is required. The users that are not authenticated to view entire data may be shown only higher level details of data that is data in generalized form whereas users like administrators can view even the lower level details. This may be managed through an inheritance hierarchy.

3. **Exaggeration and misleading:** This is an important concern to maintain security. The views that are created to support a particular query should be strictly dependent on the user involved. The user may be given an exaggerated form of data. The data that is shown may also be misleading if there is some semantic error that results in display of wrong information.
4. **Anonymity:** The anonymous is one that is not known openly to all. We also use anonymity concept in warehouses. It means that the user that is accessing the data should be anonymous to the warehouse and the warehouse should also be anonymous to the user. The first anonymity means that warehouse should not know who the user is and what level of details of data is accessed by him. The second anonymity means that user should not know from where the data is fetched, what the original source of data is and what algorithm is used in fetching of data. The connection between user and warehouse should be encrypted so that no external user may breach the inaccessible data.
5. **User profile based security:** We have already discussed user profile based security before. It means that there should be proper management of accounts of all the users. The user should not be authenticated to access data that comes under some other user and not him. One way would be to create a class of accounts. The accountants would be accessible to the data only related to accounts whereas the CEO may be given access to all the data related to sales, accounts and even the personal information of the employees. In this way the data to the unauthenticated user remains inaccessible.

7.12 BACKUP AND RECOVERY

Backup and recovery is an important aspect involved with data warehouses. This is due to the large amount of data and information stored in data warehouses. Every database, code or user directories requires a backup in case some failure occurs. Recovery stands for the recovery of the data that has been lost due to some problem.

Let us look at some of the backup definitions:

In a **complete** backup all the database files, control files and journal files are backed up.

In a **partial** backup the backup of the entire database is not taken.

Cold backup is taken when the database is shut down.

Hot backup is the backup that is not cold.

Online backup comes under hot backup category.

There are three types of recovery models:

1. Full recovery model
2. Bulk-logged recovery model
3. Simple recovery model

1. **Full recovery model:** Full recovery model is the easiest to implement. In this model the databases are required to an earlier point when they worked fine.

2. **Bulk-logged recovery model:** It gives better performance than full recovery model.. The log file required for recovery is also small in size.
3. **Simple recovery model:** This type of recovery model gives best performance. The space required for log is also minimum. This type of model suffers from the disadvantage that there is a significant loss of data in case of failure. Here the database is restored to the most recent point when the database worked fine.

Different recovery model can be chosen as per the requirement. Some model may provide with higher speed and performance but would have to give in some probability for loss of data. When selecting a model many issues may be encountered like if the operations large scale type. If there any significant loss of data in case of failure. The performance and the size of the log file is also to be considered. The complexity of the recovery model is also important.

7.12.1 Defining Backup and Recovery Strategy

The factors that need to be considered before defining backup and recovery strategy are:

1. **Data to be backed up:** Identify the size of data that is regularly required to be backed up. The metadata is also required to be backed up. The data in staging area of warehouse also requires backup.
2. **Batch Window of the Warehouse:** The backup of data requires some time. So the part of time where database is not used or the time when the minimum number of people are online is to be recognized. This is done so that it does not disappoint large number of people. However backup can also be taken online but on a certain cost. So regular backups are preferred to be taken offline.
3. **Maximum Acceptable time for recovery:** It is usual that loss of data may occur anytime without pre notice. So we require recovery techniques that are efficient. So the minimum amount of time it takes to recover the data completely and be assured that the system is working fine is to be known.
4. **Acceptable costs for recovery:** There are many backup and recovery mechanisms. A mechanism may be efficient performance wise but if it involves high cost then it would be inefficient to use that mechanism. So the acceptable costs for recovery is also to be considered.

Some other factors that need to be considered before backup are:

1. Use automated backup devices so that to eliminate problems of changing the magnetic tape manually.
2. Use incremental backups so that to eliminate the need to take backup daily.
3. The backups may also be taken at some other site from the main database site so that the backup data is not hampered if the disaster like fire occurs.
4. The backup and recovery procedure should be documented.
5. Use standard recovery format.

7.12.2 Disaster Recovery

The disasters are uncontrollable. Disasters like floods, earthquakes and fire may crop up any time. So we have to be ready with our recovery plan so that the database that has been damaged completely may be recovered. Usually when the disaster occurs entire database is damaged. The database management is partial without proper recovery techniques. The recovery of lost data may take a considerable amount of time. In some cases this time may last for days. But if it takes days then it would mean that it would cause a huge amount of load on the database. When we are starting with the planning of disaster recovery then we have to be sure of the components that are in risk. We also need to know the components that are dependent on some other components.

There would be following requirements for disaster recovery:

1. Alternate machine
2. Large size disk supporting large space.
3. Links through which communication to users may be possible.
4. Links through which communication to data sources may be possible.
5. Multiple copies of important components of software.
6. Database backup

The recovery of the lost data requires skilled and experienced staff which is capable to cope with the situation. Another important thing is that during disaster the staff also may not be present. Hence proper steps need to be taken. The recovery plan should be properly tested otherwise it would be risky to apply the plan for the first time because an inefficient plan may result in permanent loss of data or some part of data.

7.13 TUNING DATA WAREHOUSE

Proper planning is required to successfully tune the warehouse. The performance is always an important aspect. There are different tuning strategies. Some strategies are more effective than others. The warehouses that are tuned to support online transactions require different tuning strategies than decision making.

7.13.1 Tuning Data Load

We have studied the three components of data warehouse. Data is first extracted from the operational databases. The data is then transformed by transformational tools. The data is made to be compatible with warehouse. The transformation checks may occur before or after the data is loaded on the warehouse. The time when the data on the warehouse after it has been loaded must be known. When the data is loaded into the warehouse existence checks may be applied which checks if the data that is to be loaded really exists. If the existential check is not passed then it means that the information regarding that has not been loaded or the database is outdated. It never means that the data is not valid.

Figure 7.10. Data flow in a warehouse

There is a large amount of data that is required to be loaded in the warehouse. Not only the data is to be extracted from databases it is also to be transformed and checked. This creates a lot of load on the CPU because it has to check as well as move data. Parallelism is the solution to the problem. Multiple processes can be used to speed the load. To avoid bottlenecks the source data may be split into multiple files. This is done to ensure that each load has its own data. These source files may be stored at different disks. The destination may be multiple disks so that large amount of files can easily be loaded easily without contention.

7.13.2 Prioritized Tuning Steps

When tuning a warehouse different steps may be used. These steps may be prioritized for efficient tuning:

1. Tune the business rules.
2. Tune the data design.
3. Tune the Application Design.
4. Tune logical Structure of database.
5. Tune the database operations.
6. Tune the access paths.
7. Tune memory allocation.
8. Tune input/output and physical structure.
9. Tune resource contention.
10. Tune the underlying platform.

After all these steps have been performed, it would be required to assess the performance of database. Otherwise it requires tuning. The tuning is iterative process. If high performance is made in the later steps then it would be easier to make improvements in the iterative early processes.

7.13.3 Tuning Queries

There are two types of queries:

1. **Fixed queries:** Tuning is similar to tuning of relational database. The requirements are always known in advance and hence best execution plan may be carried out. The only thing that is not known is the size of query. Query is tested when desirable. Query's testing requires recording and storing most successful execution plan. To obtain an execution plan the running of query on database may not be required.

2. **Ad-hoc queries:** The number of users and what they is always affect performance. The profiles of all the users must be understood properly.

The number of users.

Are ad-hoc queries used and how oftenly?

Average size of query.

Total login time of the day.

The time when maximum users are logged in.

Number of queries run on average at a time.

Drill down to base data required or not.

If the load cannot be predicted then larger queries are required. If the number of users that are logged in is more, then more difficult the tuning task would be.

7.14 TESTING DATA WAREHOUSE

In the previous sections we studied recovery and tuning of data warehouse. How without testing of data warehouse everything stands incomplete. So effective testing of data warehouse and its components is always required. The data warehouse testing strategies will consist of a group of smaller testing strategies. A tester team is required that tests various implementation stages of data warehouse. The testing is not started at the end of implementation but it continues throughout implementation stages of the warehouse. The testing is iterative.

The various phases of database testing are as follows:

ETL	Data Load	Intial Data Load	E2E Business Testing	Mainten Ance
Business Functionality Data Quality Performance	Parametrs Settings Validation	Performance E Data Quality	UI and Interface Testing	Regressi on

Figure 7.11. Data warehouse Testing Phases

During the first phase (ETL) of testing of warehouses data quality is tested. The functionality testing is also important to validate transformation logic.

During the data load phase parameters and settings for individual modules of data load are tested and data validation is done.

In the initial data load phase data is loaded for the first time. Here performance and data quality checks are important.

In the E_2E business testing phase all the data load is complete. UI and interface testing is done during this phase.

Maintenance is the last phase. Maintenance is important in this phase. Periodic updates are applied on the data warehouse which requires temporal testing.

7.14.1 Focus Points

There are two main aspects of data warehouse testing:

1. Underlying Data
2. Data Warehouse Components

1. **Underlying Data:** It should be tested that all the data has been successfully loaded into the warehouse. It should be ensured that all the special cases are also included. Sometimes boundary cases are not covered. So it also requires testing boundary cases. Null records should also be checked. The rejected data should also be analyzed. The fields and their values in the tables must be compared to the boundary values permissible. Boundary cases should be properly tested. The data types of the fields should also be checked.

   Care must be taken during transformational logic. This is because here business requirements are translated. The data that has been transformed must be compared to the expected results. There is a technique called sampling. In this technique samples are picked up and checked against data transformations.

2. **Data warehouse components:** Performance and scalable testing is done. In this testing it is tested if the system supports additional queries and supports addition of further data. This is done to ensure that the system can be expanded. Not only scalable the system should also

give performance. The performance testing is done to ensure that there is no logical weakness in the ETL design. Data files should not be loaded multiple times and intermediate files that are not necessary should not be loaded. All the queries are also required to be tested.

The other part of component testing is orchestration testing which includes testing all the components together. It is to be ensured that all the upstream and downstream processes are working fine. This is also called integration testing. The overall architecture is to be tested. The issues like resource conflicts and deadlocks are also required to be tested. There should not be any error due to communication gap between different teams working on different components.

The other part is regression testing. It should be noted that there are periodic loads in the warehouse. The data updated in the database is required to be tested periodically. The small updation may require only small efforts whereas a bigger update may require some efforts.

Challenges during data warehouse testing are listed as follows:

1. There is huge amount of data from different sources.
2. The quality of data at data source may not be good.
3. Data verification may not be possible.
4. Entire business knowledge may not be feasible.
5. Quality of cost is high.
6. The data in sources is updated hence causing temporal inconsistency.

The best practice would be that data quality is ensured. The sampling techniques should be carried out. The best practice would be automation. This is because as we have discussed that there are periodic updates so human efforts would not be sufficient. Automate as much possible.

SUMMARY

- Aggregation is nothing but a technique to merge or consolidate dispersed data into single data. The data that has been summarized is the summarized data. Aggregation plays a key role in data warehousing.
- The warehouses are based on this historical information. It is this historical information that is used to develop a mechanism to support decision making strategies.
- Data cubes represent multi dimensional data.
- Multi dimensional data is represented by facts and dimensions.
- Dimensions are the parameters or perspectives along which data is stored.
- The common operations that could be applied to data cubes are: Roll Up, Drill Down, Slice and Dice and Pivot.
- OLAP tools are built that provides excellent interface and query processing scheme. DMQL Data Mining Query Language is used to write queries in data mining and data warehousing. Similar to RDBMS there exists DDL to define tables in warehouses.
- OLAP allows the business users and executives to access data. OLAP tools provide effective interface to analyze multi dimensional data.
- **Model:** A description of the original historical database from which it was built that can be successfully applied to new data in order to make predictions about missing values or to make statements about expected values.
- **Pattern:** A sequence of events that occurs in the database often.

- Today there a number of OLAP tools from a number of vendors are available in the market. The features and compatibility differ in some perspectives.
- OLAP functions have the capability of returning rank, row number and column function information as a scalar value in a query result.
- The chief component of OLAP is the OLAP server.
- OLAP server sits between a client and database management system.
- OLAP server understands the organization of data in the database and has special functions for analyzing the data.
- The three types of OLAP servers are ROLAP, MOLAP and HOLAP.
- Data mining is an exploratory process.
- An easy user interface is essential so that even non technical user can also use the data mining tool effectively.
- The graphic user interface should be of high quality and should be highly interactive because otherwise it would be much difficult for business users to understand it properly.
- Security is always an issue concerned to data. Various steps and strategies exist so that only authenticated users can view the aggregated warehouse data.
- Backup and recovery is an important aspect involved with data warehouses. This is due to the large amount of data and information stored in data warehouses.
- Every database, code or user directories requires a backup in case some failure occurs.
- Recovery stands for the recovery of the data that has been lost due to some problem.
- Proper planning is required to successfully tune the warehouse. The performance is always an important aspect.
- There are different tuning strategies. Some strategies are more effective than others. The warehouses that are tuned to support online transactions require different tuning strategies than decision making.
- The data warehouse testing strategies will consist of a group of smaller testing strategies.
- A tester team is required that tests various implementation stages of data warehouse.
- The testing is not started at the end of implementation but it continues throughout implementation stages of the warehouse.

REVIEW QUESTIONS

1. Explain data aggregation in terms of data warehouse.
2. What are data cubes? Explain their functionality.
3. What do you mean by query facility? Explain.
4. Write short notes on:
   (*a*) MOLAP
   (*b*) HOLAP
   (*c*) ROLAP
5. What are the phases for testing a data warehouse?

CHAPTER 8

Client-Server Computing

Client-server computing systems are comprised of two logical parts: a server that provides services and a client that requests services of the server. Together, the two form a complete computing system with a distinct division of responsibility. More technically, client-server computing relates two or more threads of execution using a consumer-producer relationship. Clients serve as the consumers in a client/server system. This means they make requests to servers for services or information and then use the response to carry out their own purpose.

The server plays the role of the producer, filling data or service requests made by clients. Client-server computing attempts to influence the capabilities of the networks used by typical corporations that are composed of many relatively powerful workstations and a limited number of dedicated servers. Client-server computing has gained popularity in the recent years due to the proliferation of low-cost hardware and the increasingly apparent truth of the theory that a model relying on monolithic applications fails when the number of users accessing a system grows too high or when too many features are integrated into a single system.

8.1 INTRODUCTION

Client-server computing is a distributed computing model in which client applications request services from server processes. Clients and servers typically run on different computers interconnected by a computer network. Any use of the Internet, such as information retrieval from the World Wide Web, is an example of client-server computing. However, the term is generally applied to the systems in which an organization runs programs with multiple components distributed among computers in a network. The concept is frequently associated with enterprise computing, which makes the computing resources of an organization available to every part of its operation.

A client application is a process or program that sends messages to a server via the network. Those messages request the server to perform a specific task, such as looking up a customer record in a database or returning a portion of a file on the server's hard disk. The client manages local resources such as a display, keyboard, local disks, and other peripherals. The server process or program listens for client requests that are transmitted via the network. Servers receive those requests and perform actions such as database queries and reading files. Server processes typically run on powerful PCs, workstations, or mainframe computers.

An example of a client-server system is a banking application that allows a clerk to access account information on a central database server. All access is done via a PC client that provides a graphical user interface (GUI). An account number can be entered into the GUI along with how much money is to be withdrawn or deposited, respectively. The PC client validates the data provided by the clerk, transmits the data to the database server, and displays the results that are returned by the server.

The client-server model is an extension of the object based (or modular) programming model, where large pieces of software are structured into smaller components that have well defined interfaces. This decentralized approach helps to make complex programs maintainable and extensible. Components interact by exchanging messages or by Remote Procedure Calling. The calling component becomes the client and the called component becomes the server.

A client-server environment may use a variety of operating systems and hardware from multiple vendors; standard network protocols like TCP/IP provide compatibility. Vendor independence and freedom of choice are further advantages of the model. Client-server systems can be scaled up in size more readily than centralized solutions since server functions can be distributed across more and more server computers as the number of clients increases. Server processes can thus run in parallel, each process serving its own set of clients. However, when there are multiple servers that update information, there must be some coordination mechanism to avoid inconsistencies.

The drawbacks of the client-server model are that security is more difficult to ensure in a distributed environment than it is in a centralized one, that the administration of distributed equipment can be much more expensive than the maintenance of a centralized system, that data distributed across servers needs to be kept consistent, and that the failure of one server can render a large client–server system unavailable. If a server fails, none of its clients can make further progress, unless the system is designed to be fault-tolerant.

The computer network can also become a performance or reliability bottleneck: if the network fails, all servers become unreachable. If one client produces high network traffic then all clients may suffer from long response times.

Client-server is an architecture in which a system's functionality and its processing are divided between the client PC and a database server. System functionality, such as programming logic, business rules and data management is segregated between the client and server.

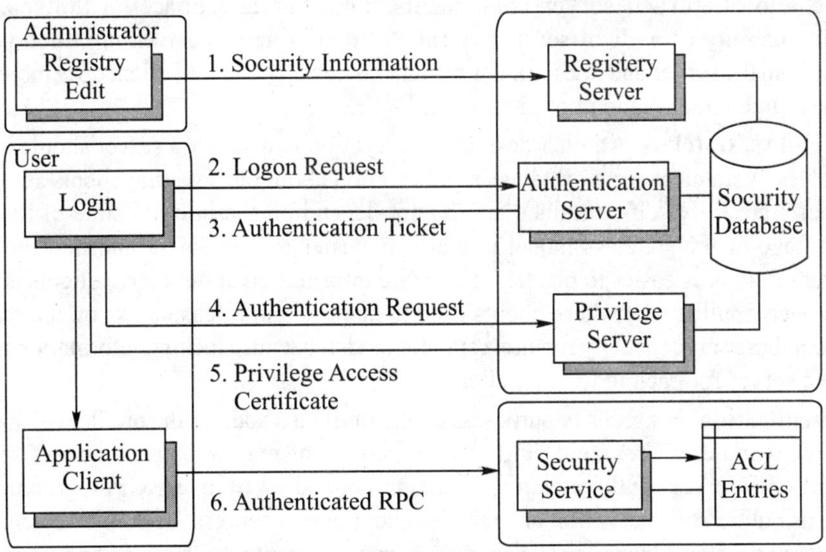

Figure 1. Client-Server System

Client-server computing comprises three building blocks:

- The client
- The server (may be more than one)
- The network (tying the client and server together).

A logical and physical separation exists between the client and server and the client/server system co-ordinates the work of both of these components and efficiently uses each one's available resources to complete assigned tasks. This separation of client and server provides an open and flexible environment where mix and match of hardware and operating systems is the rule. The network ties everything together. Today the client applications run predominantly on PCs connected to a network (or LAN). The servers are also connected to the network and know how to connect to their clients.

Design Considerations

An important design consideration for large client-server systems is whether a client talks directly to the server, or whether an intermediary process is introduced between the client and the server. The former is two-tier architecture; the latter is three-tier architecture.

The two-tier architecture is easier to implement and is typically used in small environments (one or two servers with one or two dozens of clients). However, two-tier architecture is less scalable than three-tier architecture.

In the three-tier architecture, the intermediate process is used for decoupling clients and servers. The intermediary can cache frequently used server data to ensure better performance and scalability. Performance can be further increased by having the intermediate process distribute client requests to several servers so that requests execute in parallel.

The intermediary can also act as a translation service by converting requests and replies from one format to another or as a security service that grants server access only to trusted clients. Other important design considerations are:

- **Fat vs. thin client:** A client may implement anything from a simple data entry form to a complex business application. An important design consideration is how to partition application logic into client and server components. This has an impact on the scalability and maintainability of a client-server system. A "thin" client receives information in its final form from the server and does little or no data processing. A "fat" client does more processing, thereby lightening the load on the server.
- **Stateful vs. stateless:** Another design consideration is whether a server should be stateful or stateless. A stateless server retains no information about the data that clients are using. Client requests are fully self-contained and do not depend on the internal state of the server. The advantage of the stateless model is that it is easier to implement and that the failure of a server or client is easier to handle, as no state information about active clients is maintained. However, applications where clients need to acquire and release locks on the records stored at a database server usually require a stateful model, because locking information is maintained by the server for each individual client.
- **Authentication:** For security purposes servers must also address the problem of authentication. In a networked environment, an unauthorized client may attempt to access sensitive data stored on a server. Authentication of clients is handled by using cryptographic techniques such as public key encryption or special authentication servers. In public key encryption, the client application "signs" requests with its private cryptographic key, and encrypts the data in the request with a secret session key known only to the server and to the client. On receipt

of the request, the server validates the signature of the client and decrypts the request only if the client is authorized to access the server.

- **Fault tolerance:** Applications such as flight-reservation systems and real-time market data feeds must be fault-tolerant. This means that important services remain available in spite of the failure of part of the computer system on which the servers are running (high availability), and that no information is lost or corrupted when a failure occurs (consistency). For the sake of high availability, critical servers can be replicated, which means they are provided redundantly on multiple computers. If one replica fails then the other replicas still remain accessible by the clients. To ensure consistent modification of database records stored on multiple servers, a transaction processing monitor can be installed. TP monitors are intermediate processes that specialize in managing client requests across multiple servers. The TP monitor ensures that such requests happen in an "all-or-nothing" fashion and that all servers involved in such requests are left in a consistent state, in spite of failures.

Distributed Object Computing

Distributed object computing (DOC) is a generalization of the client-server model. Object-oriented modeling and programming are applied to the development of client-server systems. Objects are pieces of software that encapsulate an internal state and make it accessible through a well-defined interface. In DOC, the interface consists of object operations and attributes that are remotely accessible. Client applications may connect to a remote instance of the interface with the help of a naming service. Finally, the clients invoke the operations on the remote object. The remote object thus acts as a server. This use of objects naturally accommodates heterogeneity and autonomy. It supports heterogeneity since requests sent to server objects depend only on their interfaces and not on their internals. It permits autonomy because object implementations can change transparently provided they maintain their interfaces.

If complex client-server systems are to be assembled out of objects, then objects must be compatible. Client-server objects have to interact with each other even if they are written in different programming languages and run on different hardware and operating system platforms.

Standards are required for objects to interoperate in heterogeneous environments. One of the widely adopted vendor-independent DOC standards is the OMG (Object Management Group) CORBA (Common Object Request Broker Architecture) specification.

CORBA consists of the following building blocks:

- **Interface Definition Language:** Object interfaces are described in a language called IDL (Interface Definition Language). IDL is a purely declarative language resembling Cþþ. It provides the notion of interfaces (similar to classes), of interface inheritance, of operations with input and output arguments, and of data types that can be passed along with an operation. IDL serves for declaring remotely accessible server objects in a platform-and programming language-neutral manner, but not for implementing those objects. CORBA objects are implemented in widely used languages such as Cþþ, C, Java, and Smalltalk.
- **Object Request Broker:** The purpose of the ORB (Object Request Broker) is to find the server object for a client request, to prepare the object to receive the request, to transmit the request from the client to the server object, and to return output arguments back to the client application. The ORB mainly provides an object-oriented RPC facility.
- **Basic Object Adapter:** The BOA (Basic Object Adapter) is the primary interface used by a server object to gain access to ORB functions. The BOA exports operations to create object references, to register and activate server objects, and to authenticate requests. An object reference is a data structure that denotes a server object in a network. A server installs its

reference in a name server so that a client application can retrieve the reference and invoke the server. The object reference provides the same interface as the server object that it represents. Details of the underlying communication infrastructure are hidden from the client.

- **Dynamic Invocation Interface:** The DII (Dynamic Invocation Interface) defines functions for creating request messages and for delivering them to server objects. The DII is a low-level equivalent of the communication stubs (message-passing interfaces) that are generated from an IDL declaration.
- **Internet Inter-ORB Protocol:** The Internet Inter-ORB Protocol (IIOP) allows CORBA ORBs from different vendors to interoperate via a TCP/IP connection. IIOP is a simplified RPC protocol used to invoke server objects via the Internet in a portable and efficient manner.
- **Interface and Implementation Repository:** The CORBA Interface Repository is a database containing type information (interface names, interface operations, and argument types) for the interfaces available in a CORBA system. This information is used for dynamic invocation via the DII, for revision control, and so forth. The Implementation Repository provides information allowing an ORB to locate and launch server objects.

8.2 TRADITIONAL HOST SYSTEM

Several years ago, many computing environments consisted of mainframes hooked to dumb terminals that only did processing at the mainframe. Over the years, personal computers started to replace these dumb terminals but the processing continued to be done on the mainframe. The improved capacity of personal computers were largely ignored or used on an individual level. With so much computing power idle, many organizations started thinking about sharing, or splitting, some of the processing demands between the mainframe and the PC. Client-server technology evolved out of this movement for greater computing control and more computing value.

Client-server refers to the way in which software components interact to form a system that can be designed for multiple users. This technology is a computing architecture that forms a composite system allowing distributed computation, analysis, and presentation between PCs and one or more larger computers on a network. Each function of an application resides on the computer most capable of managing that particular function. There is no requirement that the client and server must reside on the same machine. In practice, it is quite common to place a server at one site in a local area network (LAN) and the clients at the other sites. The client, a PC or workstation, is the requesting machine and the server, a LAN file server, mini or mainframe, is the supplying machine. Clients may be running on heterogeneous operating systems and networks to make queries to the server(s).

Networks provide connectivity between client-server and the protocols that they use to communicate. The Internet provides connectivity between systems that function as clients, servers, or both. Many services used on the Internet are based on client-server computing model. File Transfer Protocol (FTP) for example uses client-server interactions to exchange files between systems. An FTP client requests a file that resides on another system. An FTP server on the system where the file resides handles the client's request. The server gets access to the file and sends the file back to the client's system. Market researchers have projected enormous growth in the client-server area. This growth seems to have come at the expense of the mainframe market, which has stagnated. While the movement towards migrating from the mainframe to client-server architecture is gaining momentum, there are several distinct drawbacks since most of the client-server tools and methodologies are not in place and the associated administration support is still undefined.

Local Terminals are responsible for display and keyboard for user input and viewing capabilities. Local Terminals do not contain any intelligent processing capabilities.

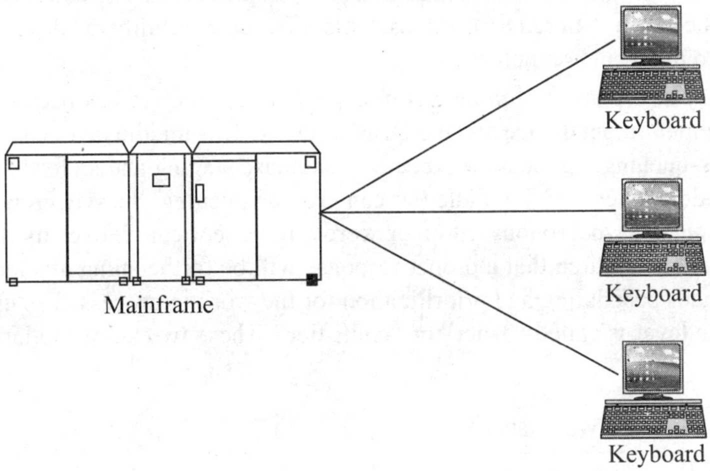

Figure 8.2. Non-client server system

8.3 2-TIER ARCHITECTURE

In a 2-tier client-server architecture, RPC's or SQL are typically used to communicate between the client and server. The server is likely to have support for stored procedures and triggers. These mean that the server can be programmed to implement business rules that are better suited to run on the server than the client, resulting in a much more efficient overall system.

Since 1992 software vendors have developed and brought to market many toolsets to simplify development of applications for the 2-tier client-server architecture. The best known of these tools are Microsoft's Visual Basic, Borland's Delphi, and Sybase's PowerBuilder. These modern, powerful tools combined with literally millions of developers who know how to use them, means that the 2-tiered client/server approach is a good and economical solution for certain classes of problems.

The 2-tiered client-server architecture has proven to be very effective in solving workgroup problems. "Workgroup", is loosely defined as a dozen to 100 people interacting on a LAN. For bigger enterprise-class problems and/or applications that are distributed over a WAN, use of this 2-tier approach has generated some problems.

What typically happens with client/server in large enterprise environments is that the performance of 2-tier architecture deteriorates as the number of on-line users increases. The reason for this is due to the connection process of the DBMS server. The DBMS maintains a thread for each client connected to the server. Even when no work is being done, the client and server exchange "keep alive" messages on a continuous basis. If something happens to the connection, the client must go through a session reinitiating process. With 50 clients and today's typical PC hardware, this is no problem. When one has 2,000 clients on a single server, however, the resulting performance isn't likely to be satisfactory.

The data language used to implement server procedures in SQL server type data base management systems is proprietary to each vendor. Oracle, Sybase, Informix and IBM, for example, have implemented different language extensions for these functions. Proprietary approaches are fine from a performance point of view, but are a disadvantage for users who wish to maintain flexibility and choice in which DBMS is used with their applications.

Another problem with the 2-tiered approach is that current implementations provide no flexibility in "after the fact partitioning". Once an application is developed it isn't easy to move or split some of the program functionality from one server to another. This would require manually regenerating procedural code. In some of the newer 3-tiered approaches, tools offer the capability to "drag and drop" application code modules onto different computers.

The industry's response to limitations in the 2-tier architecture has been to add a third, middle tier, between the input/output device and the DBMS server. This middle layer can perform a number of different functions-queuing, application execution, database staging and so forth. Just to illustrate one advantage of a middle layer, if that middle tier can provide queuing, the synchronous process of the 2-tier approach becomes asynchronous. In other words, the client can deliver its request to the middle layer, disengage and be assured that a proper response will be forthcoming at a later time. In addition, the middle layer adds scheduling and prioritization for the work in process. The use of an architecture with such a middle layer is called "3-tier" or "multi-tier". These two terms are largely synonymous in this context.

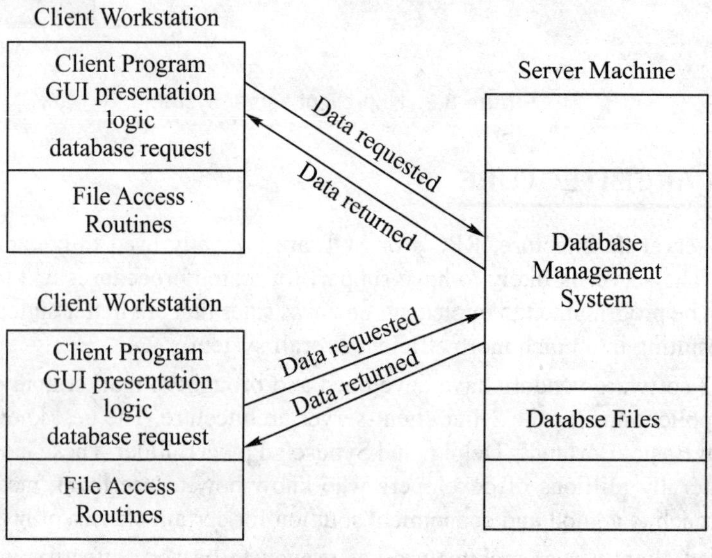

Figure 8.3. 2-Tier Architecture

Advantages and Disadvantages of 2-Tier Architecture

Advantages	*Disadvantages*
Development Issues: • Simple structure • Easy to setup and maintain	*Development Issues:* • Complex application rules difficult to implement in database server-requires more code for the client • Complex application rules difficult to implement in client and have poor performance • Changes to business logic not automatically enforced by a server-changes require new client side software to be distributed and installed • Not portable to other database server platforms

(*Contd.*.)

Performance:	*Performance:*
• Adequate performance for low to medium volume environments. • Business logic and database are physically close, which provides higher perfomance.	• Inadequate performance for medium to high volume environments, since database server is required to perform business logic. This slows down database operations on database server.

8.4 3-TIER ARCHITECTURE

The three-tier client/server architecture is an evolution of the traditional two-tier model, and is receiving increased interest, particularly for large business applications. The main difference is that in a three-tier architecture, most of the functionality is separated out in a middle layer, called *application servers*, as shown in figure 8.4. Also, each client can use several application servers, each of which in turn can work against several databases.

Figure 8.4. 2-Tier Architecture

The most basic type of middle layer is the transaction processing monitor or TP monitor. One can think of a TP monitor as a kind of message queuing service. The client connects to the TP monitor instead of the database server. The transaction is accepted by the monitor, which queues it and then takes responsibility for managing it to correct completion.

TP monitors first became popular in the 1970's on mainframes. On-line access to mainframes was available through one of two metaphors-time sharing or transaction processing (OLTP). Time sharing was used for program development and the computer's resources were allocated with a simple scheduling algorithm like round robin. OLTP scheduling was more sophisticated and priority driven. TP monitors were almost always used in this environment, and the most popular of these was IBM's CICS (pronounced "kicks").

As client-server applications gained popularity over the early 1990's, the use of TP monitors dropped by the wayside. That happened principally because many of the services provided by a TP monitor were available as part of the DBMS or middleware software provided by vendors like Sybase, Gupta, and Oracle. Those embedded (in the DBMS) TP services have acquired the nickname "TP Lite". The "Lite" term comes from experience that DBMS-based transaction processing works OK as long as a relatively small number (<100) of clients are connected.

TP monitors (TP Heavy) have staged a comeback because their queuing engines provide a funneling effect, reducing the number of threads a DBMS server needs to maintain. The client connects with the monitor, which accepts the message and queues it for processing against the database. Once the monitor has accepted the message, the client can be released for further processing. The synchronous session based computing of a 2-tier architecture, then, becomes asynchronous through the insertion of the TP monitor into the equation. The monitor smoothes out and lowers the overhead of accessing the database server.

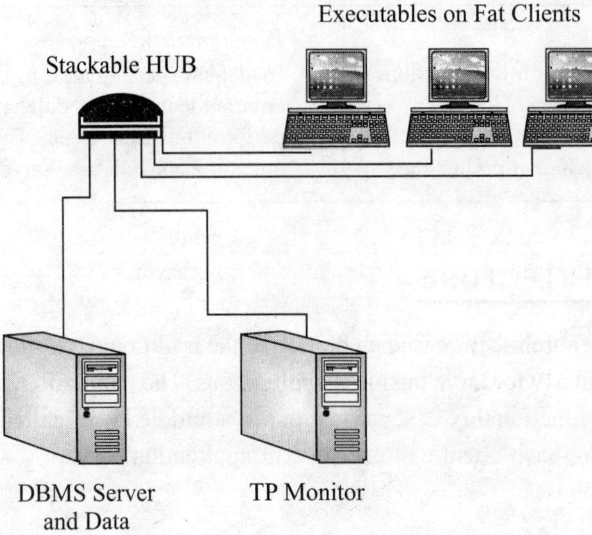

Figure 8.5. 3-Tier TP Monitor

Some other key services a monitor provides are:

- the ability to update multiple different DBMS in a single transaction;
- connectivity to a variety of data sources including flat files, non relational DBMS, and the mainframe;
- the ability to attach priorities to transactions;
- and robust security, including Kerberos.

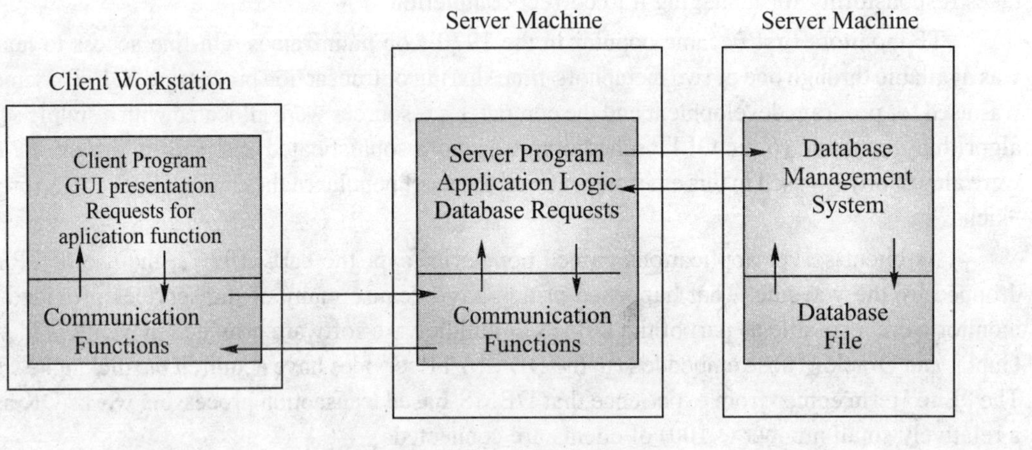

Figure 8.6. 3-Tier Architecture

The net result of using 3-tier client/server architecture with a TP monitor is that the resulting environment is *FAR* more scalable than a 2-tier approach with direct client to server connection. For really large (*e.g.*, 1,000 user) applications, a TP monitor is one of the most effective solutions.

As one might expect, there is a downside to network-based TP monitors. At this point in time, the major problem with using this approach is that the code to implement TP monitors is usually written in a lower level language (like COBOL), and support for TP monitors is not (yet) widely available in the most popular visual toolsets like PowerBuilder or Visual Basic.

3-Tier with a Messaging Server

Messaging provides still another technology to implement 3-tier computing. It is available today from companies such as IBM, DEC, Sybase, and Oracle. A messaging server can be thought of as a kind of "second generation" TP monitor and provides the same funneling process. Messages are processed asynchronously with the appropriate priority level. And, like a TP monitor, a messaging server provides connectivity to data sources other than RDBMS.

A message is a self contained object that carries information about what it is, where it needs to go, and what should happen when it reaches its destination. There are at least two parts to every message; the header contains priority, and address and an ID number. The body of the message contains the information being sent, which can be anything including text, images or transactions.

A primary difference from TP Monitors is that a message server architecture is designed around intelligence in the message itself as opposed to a TP monitor environment which places the system intelligence in the monitor or the process logic of the application server.

In a TP monitor environment the transactions are simply dumb packets of data. They travel over a pre-existing and pre-defined connection to the TP Monitor. The TP Monitor interrogates and processes the transaction, usually submitting the request to a server tier application. If the TP Monitor doesn't understand the data, it doesn't get processed. Ultimately, the TP Monitor needs to know as much about the transaction as the server tier does.

Contrasting with this, in a message-based architecture there's intelligence in the message itself. The message server just becomes a container of messages and their stored procedures. The operations performed by the message server on the message are communications related (*e.g.*, encrypt message over one service and decrypt message sent over another service). For the most part, messages are treated as discrete objects. The message contains all the information needed to transverse network services (*i.e.*, network addresses, both logical and physical). Because the message contains the intelligence, the middle tier of a message-based system is more flexible than a TP monitor. For one kind of message, the middle tier may simply serve as a routing point between two kinds of network services. For another kind of message, the middle tier may execute a stored procedure or business rule as directed by the message. This abstraction of the middle-tier away from the contents and behavior of the information flowing through it makes the system more portable to different environments and networks. The specifics of communicating the information are hidden underneath the messaging service.

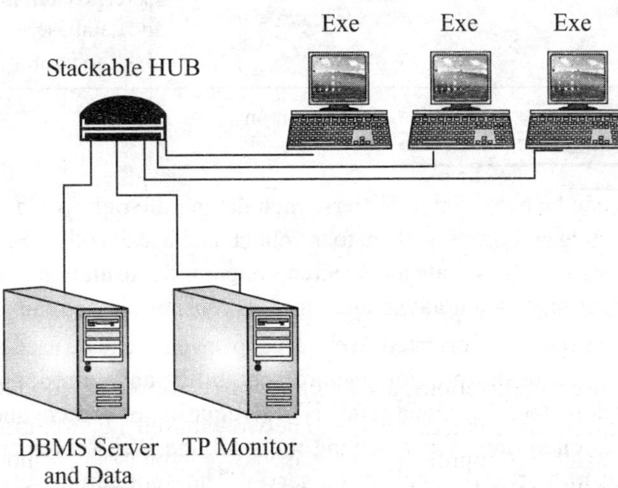

Figure 8.7. Message based Client-Server

Messaging systems are designed for robustness. By using store and forward logic, they provide message delivery after and around failures. They also provide independence from the enabling technologies such as wired or wireless or protocols. They don't require a persistent connection between the client and server. They are robust because message delivery can be programmed to occur after or around failures. Because messaging systems support an emerging wireless infrastructure, they should become popular for supporting mobile and occasionally connected workers.

A typical message server architecture would look like the figure 8.7, which looks just like any of the other 3-tier approaches we're going to discuss. The architecture of an application that uses messaging services will turn out to look similar to an approach that depends on distributed objects and object request brokers (ORB's) for communication. If one is unwilling or unable to wait for the arrival of distributed object technologies to build your application (widespread popularity probably won't happen with ORB's until the end of the 1990's), one can construct a reasonable clone using the messaging approaches that are now available. When distributed objects are a reality, one can migrate its application if that seems like the best move.

Advantages and Disadvantages of 3-Tier Architecture

Advantages	*Disadvantages*
Development Issues: • Complex application rules easy to implement in application server • Business logic off-loaded from database server and client, which improves performance • Changes to business logic automatically enforced by server - charges require only new application server software to be installed • Application server logic is portable to other database server platforms by virtue of the application software	*Development Issues:* • More complex structure • More difficult to setup and maintain.
Performance: • Superior performance for medium to high volume environments	*Performance:* • The physical separation of application servers containing business logic functions and database servers containing databases may moderately affect performance.

8.4 N-TIER ARCHITECTURES

The 3-tier architecture can be extended to N-tiers when the middle tier provides connections to various types of services, integrating and coupling them to the client, and to each other. Partitioning the application logic among various hosts can also create an N-tiered system. Encapsulation of distributed functionality in such a manner provides significant advantages such as reusability, and thus reliability.

As applications become Web-oriented, Web server front ends can be used to offload the networking required to service user requests, providing more scalability and introducing points of functional optimization. In this architecture, the client sends HTTP requests for content and presents the responses provided by the application system. On receiving requests, the Web server either returns the content directly or passes it on to a specific application server. The application server might then run CGI scripts for dynamic content, parse database requests, or assemble formatted responses to client queries, accessing dates or files as needed from a back-end database server or a file server.

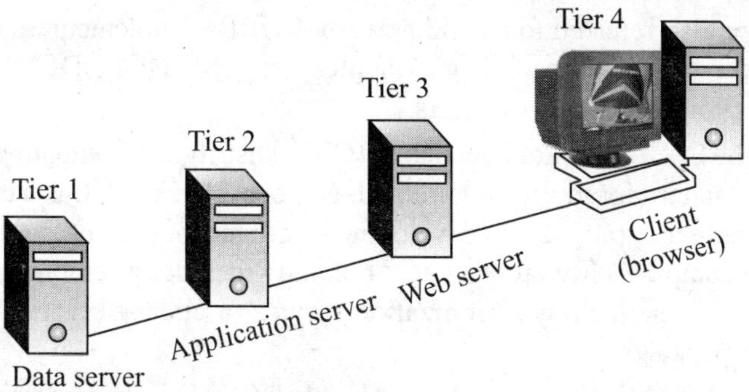

Figure 8.8. Web-Oriented N-Tiered Architecture

By segregating each function, system bottlenecks can be more easily identified and cleared by scaling the particular layer that is causing the bottleneck. For example, if the Web server layer is the bottleneck, multiple Web servers can be deployed, with an appropriate server load-balancing solution to ensure effective load balancing across the servers.

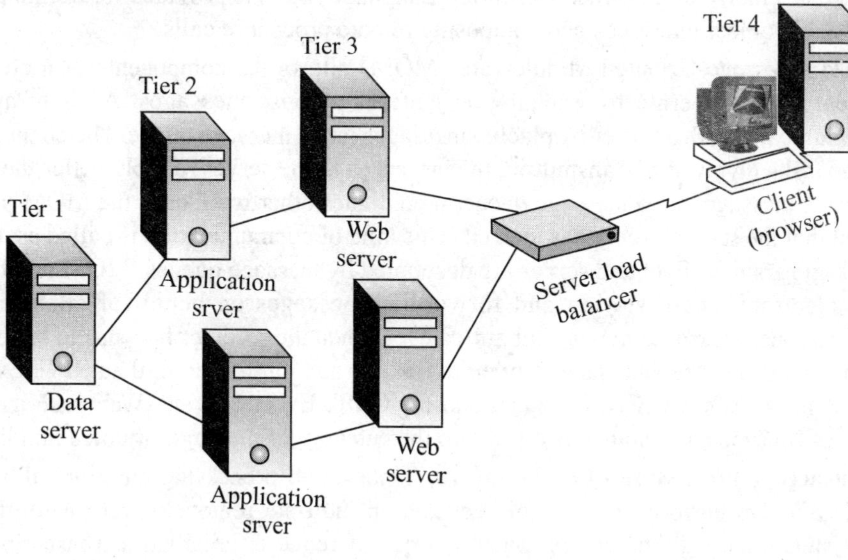

Figure 8.9. Four-Tiered Architecture with Server Load Balancing

The N-tiered approach has several benefits:

(*a*) Different aspects of the application can be developed and rolled out independently.

(*b*) Servers can be optimized separately for database and application server functions.

(*c*) Servers can be sized appropriately for the requirements of each tier of the architecture.

(*d*) More overall server horsepower can be deployed.

8.6 CLIENT-SERVER TOOLKIT

A wide range of software toolkits for building client-server software is available on the market today. Client-server toolkits are also referred to as middleware. CORBA implementations are an example of well-known client-server middleware. Other examples are OSF DCE, DCOM, message-oriented middleware, and transaction processing monitors.

- **OSF DCE:** The Open Software Foundation (OSF) Distributed Computing Environment (DCE) is a de facto standard for multivendor client–server systems. DCE is a collection of tools and services that help programmers in developing heterogeneous client–server applications. DCE is a large and complex software package; it mainly includes a remote procedure call facility, a naming service, a clock synchronization service, a client–server security infrastructure, and a threads package.
- **DCOM:** Distributed Component Object Model (DCOM) is Microsoft's object protocol that enables ActiveX components to communicate with each other across a computer network. An ActiveX component is a remote accessible object that has a well defined interface and is self-contained. ActiveX components can be embedded into Web documents, so that they download to the client automatically to execute in the client's Web browser. DCOM provides a remote instantiation facility allowing clients to create remote server objects. It also provides a security model to let programmers restrict who may create a server object and who may invoke it. Finally, an Interface Definition Language (IDL) is provided for defining remotely accessible object interfaces and composing remote procedure calls.
- **MOM:** Message-Oriented Middleware (MOM) allows the components of a client–server system to interoperate by exchanging general purpose messages. A client application communicates with a server by placing messages into a message queue. The client is relieved of the tasks involved in transmitting the messages to the server reliably. After the client has placed a message into a message queue, it continues other work until the MOM informs the client that the server's reply has arrived. This kind of communication is called asynchronous messaging, since client and server are decoupled by message queues. MOM functions much like electronic mail, storing and forwarding messages on behalf of client and server applications. Messages may be submitted even when the receiver happens to be temporarily unavailable, and are thus inherently more flexible and fault-tolerant than RPC. Examples of MOM are IBM's MQSeries product and the OMG Event Service. Web push technologies such as Marimba's Castanet also fall into the category of message-oriented middleware.
- **Transaction Processing (TP) Monitors:** Transaction processing monitors allow a client application to perform a series of requests on multiple remote servers while preserving consistency among the servers. Such a series of requests is called a transaction. The TP monitor ensures that either all requests that are part of a transaction succeed, or that the servers are rolled back to the state they had before the unsuccessful transaction was started. A transaction fails when one of the involved computers or applications goes down, or when any of the applications decides to abort the transaction.

A TP monitor can be used within a banking system when funds are withdrawn from an account on one database server and deposited in an account on another database server. The monitor makes sure that the transaction occurs in an "all or nothing" fashion. If any of the servers fails during the transfer then the transaction is rolled back such that both accounts are in the state they were before transaction was started.

8.7 BENEFITS OF THE CLIENT/SERVER MODEL

- Divides Application processing across multiple machines: Non-critical data and functions are processed on the client; Critical functions are processed on the server
- Optimizes Client Workstations for data input and presentation (*e.g.*, graphics and mouse support)
- Optimizes the Server for data processing and storage (*e.g.*, large amount of memory and disk space)
- Scales Horizontally – Multiple servers, each server having capabilities and processing power, can be added to distribute processing load.
- Scales Vertically – Can be moved to more powerful machines, such as minicomputer or a mainframe to take advantage of the larger system's performance
- Reduces Data Replication - Data stored on the servers instead of each client, reducing the amount of data replication for the application.

8.8 THIN AND THICK CLIENT SERVER COMPUTING

The TCSC model provides global integration that can significantly lower the cost of ownership to customers looking for mission-critical, centralized server-based computing with maximum security, standardized desktop deployment, and 100% remote client management tools. Here is a list of the major benefits offered by thin clients:

- Rapid device/application deployment.
- Application upgrades are deployed to a single server instead of hundreds or thousands of individual PCs.
- Because individual thin clients can be configured from a central location, it takes minutes instead of hours to deploy a new client.
- Simplified IT management
- Applications are typically deployed on the server and merely accessed by the thin client, which can dramatically reduce the amount of time the administrator must spend on client management.
- Data is automatically backed up for users.
- Users cannot access or corrupt system files.
- Network administrators can manage clients remotely.
- Due to the relatively small client images used on thin clients, complete remote reimaging becomes a viable solution to a local client problem. This solution may reduce the amount of time an administrator spends diagnosing problems.
- Increased reliability and security.
- Fewer maintenance events and less down time due to no moving parts to fail.
- Fewer conflicts and configuration issues and reduced threat of viruses or data theft as a result of no removable media.
- Because data is stored on network storage, the potential of data loss due to a hard drive crash is reduced.

In the world of client/server architecture, one needs to determine if it will be the client or the server that handles the bulk of the workload. By client, we mean the application that runs on a personal computer or workstation and relies on a server to perform some operations. Thick or thin client architecture is actually quite similar. In both cases, one can consider it as being the client application running on a PC whose function is to send and receive data over the network to the server program. The server would normally communicate that information to the middle-tier software (the backend), which retrieves and stores that information from a database.

While they share similarities, there are many differences between thick and thin clients. Thick and thin are the terms used to refer to the hardware (*e.g.*, how a PC communicates with the server), but the terms are also used to describe applications.

Thin Clients

A thin client is designed to be especially small so that the bulk of the data processing occurs on the server. Although the term thin client often refers to software, it is increasingly used for the computers, such as network computers and Net PCs that are designed to serve as the clients for client/server architectures. A thin client is a network computer without a hard disk drive. They act as a simple terminal to the server and require constant communication with the server as well. Thin clients provide a desktop experience in environments where the end user has a well-defined and regular number of tasks for which the system is used. Thin clients can be found in medical offices, airline ticketing, schools, governments, manufacturing plants and even call centers. Along with being easy to install, thin clients also offer a lower total cost of ownership over thick clients.

Thick Clients

In contrast, a thick client (also called a fat client) is one that will perform the bulk of the processing in client/server applications. With thick clients, there is no need for continuous server communications as it is mainly communicating archival storage information to the server. As in the case of a thin client, the term is often used to refer to software, but again is also used to describe the networked computer itself. If the applications require multimedia components or that is bandwidth intensive, one will also want to consider going with thick clients. One of the biggest advantages of thick clients rests in the nature of some operating systems and software being unable to run on thin clients. Thick clients can handle these as it has its own resources.

Thick vs. Thin Clients—A Quick Comparison

Thin Clients	*Thick Clients*
Easy to deploy as they require no extra or specialized software installation.	More expensive to deploy and more work for IT to deploy.
Needs to validate with the server after data capture.	Data verified by client not server (immediate validation).
If the server goes down, data collection is halted as the client needs constant communication with the server.	Robust technology provides better uptime.
Cannot be interfaced with other equipment (in plants or factory settings for example).	Only needs intermittent communication with server.
Clients run only and exactly as specified by the server.	More expensive to deploy and more work for IT to deploy.

More downtime.	Require more resources but less servers.
Portability in that all applications are on the server so any workstation can access.	Can store local files and applications.
Opportunity to use older, outdated PCs as clients.	Reduced server demands.
Reduced security threat.	Increased security issues.

8.9 SECURITY IN CLIENT SERVER SYSTEMS

Security is often not given the consideration it requires in client/server partly because the implementation of security represents a cost that does not reflect an immediate return and partly because purchasers are often not aware of security issues and buy the cheapest client because they can get more for their dollar.

Security in Client/Server

The distribution of services in client/ server increases the susceptibility of these systems to damage from viruses, fraud, physical damage and misuse than in any centralized computer system. With businesses moving towards multi-vendor systems, often chosen on the basis of cost alone, the security issues multiply. Security has to encompass the host system, PCs, LANs, workstations, global WANs and the users.

However, every level of system security requires dollars and additional steps for the users. The cost and inconvenience (to the users) associated with security must be balanced against the cost and inconvenience of corrupted or insecure data.

The Client

As the primary front end device of a client/server system the desktop is typically the least mature with respect to security. Where the mainframe environment had a period of steady growth for more than 20 years, with the demand for security products that went with that growth, client/server has increased in popularity only since the early 1990s, and its hardware components, including the PC, has not been subject to the same security considerations that the mainframe has.

The client machines pose a threat to security as they can connect to the servers, and access their data, that are elsewhere in an organization. One large problem is that they are easily accessible and easy to use. They are usually located in open plan offices that present a pleasant environment for users (and intruders) making it impossible to lock them away when unattended. Products are available that offer a measure of physical security by locking or bolting equipment and cabling into place.

Physical protection for the client machines can include disk drive locks, or even diskless workstations to prevent the loading of unauthorized software and viruses. The cases can be fitted with locks to prevent access to the hard drives, and memory. One of the greatest risks with the client workstations is that the operating system is easily and directly accessible to the end user which exposes the whole system to a number of risks. The workstation operating system assumes that the person who turns it on is the owner of all files on the computer, including the configuration files. Even if the client/ server application has good security, that security might not be able to counteract attacks at the operating system level which could corrupt data passed to other tiers of the client/server system. The tighter security now being offered with Windows NT addresses some of these issues.

Often client machines are bought based on cost alone, but in this case the cheapest may actually cost the company more as additional security features are added. Weak cases have to be strengthened to prevent opening and supplied operating systems replaced with more robust ones, for example the login password can be bypassed on Windows 95 machines whereas Windows NT V4.0 prevents access to the file system until authenticated but it costs two to three times more.

The Network

The network connecting clients and servers is a less than secure vehicle that intruders can use to break into computer systems and their various resources. Using publicly available utilities and hardware an attacker can eavesdrop on a network, or "sniff" the network to read packets of information. These packets can contain useful information, *e.g.,* passwords, company details, etc, or reveal weaknesses in the system that can be used to break into the system.

Encryption of data can solve the problem of attackers sniffing the network for valuable data. Encryption involves converting the readable data into unreadable data. Only those knowing the decryption key can read the data. A problem here is that some network operating systems don't start encryption until the user has been authenticated (*i.e.* the password is sent unencrypted).

Most systems employ re-usable passwords for authenticating users which allows an attacker to monitor the network, extract the login information and access the system posing as that user. Even if the password is encrypted the intruder can just inject that packet into the network and gain access. The problem is compounded when, to maintain that single system illusion, only one login is required to access all servers on the network. Customers want a "single system image" of all networked computing resources, in which all systems management and administration can be handled within a single pool of system resources.

To have a secure network it must conform to four basic principles of a trusted computing base (TCB):

- Identification and authorization
- Discretionary control
- Audit
- Object reuse.

The fundamental aspect of the TCB is that if you can trust all of the security features, then the network can also be trusted. The TCB must be self protected against tampering and malicious, inadvertent altering and attempts to circumvent it.

The National Computer Security Centre's (NCSC) evaluation model specifies a C2 level of security that provides the above features. C2 is a defined level for operating systems requiring users and applications to be authenticated before gaining access to any operating system resource. All clients must provide an authenticated user ID, access control lists must protect all resources, audit trails must be provided, and access rights must not be passed to other users that re-use the same items.

User authentication can be managed by the Kerberos authentication mechanism. The Kerberos protocol, with add ons introduced by the Open Software Foundations Distributed Computing Environment (OSF DCE), fulfills the authentication requirement of C2. Kerberos is a secret key network authentication system that uses DES for encryption and authentication. It authenticates every user for every application. It consists of three distinct services that provide access to network resources, Registry, Authentication and Privilege Servers.

In DCE, users, servers and client computers are all referred to as "principals". The Registry server creates user accounts for the network principals and unique user IDs are created and stored with other information in the security database. At the time of logon, the DCE security system uses encrypted "keys" that reflect private validation information, such as a principal's password, the server and services or data required. This process of validating a principal through credentials is known as authentication. At the time of the logon request, a logon request is submitted to the Authentication Server.

The Authentication Server responds to the logon service by forwarding a ticket that is encrypted using the user's secure key. Once received by the logon service, an attempt to decrypt the ticket is made using the password supplied by the principal. If a valid password has been supplied, a new ticket will be created by the logon service. This ticket is then forwarded to a Privilege Server, where a new ticket is created that will be used to provide access to a specific application server and the specific services required. Properties of the Kerberos tickets include encryption of the ticket to insure its authenticity. If any data within the ticket is modified, the ticket becomes invalid. Further, tickets are issued with a specific time duration to protect against an attempt to modify, copy, or utilize someone else's ticket. The DCE Security Service is designed to allow applications to authenticate not only users, but servers as well. This is provided with each application server having again its own secret key that can be used by the security service to send packages that only the true server can decrypt.

Once the client receives the ticket, it presents the ticket and request to the targeted server. The server then compares the incoming ticket with the ticket from the Kerberos server to verify that the client has access.

Applications, files, and other DCE services then use Access Control Lists (ACLs) to identify specific access privileges of a principal, whose individual or group identity is provided via the ticket. An ACL might identify whether a principal has read, write, test, or other permissions against a file or records within a database. If the ACL permissions match the action requested, the access is granted; otherwise, the RPC call is denied and a result returned to the calling application.

The purpose of the DCE Security Service is to provide a security architecture that provides both users and servers the ability to communicate in a manner which is very difficult to impersonate. This allows servers to offer service only to authorised users, and it allows users to have confidence that they are communicating with the "real" server.

Audit services allow network managers to monitor user activities, including attempted logons and the file servers or files used. It is achieved by monitoring all user workstations and recording transaction activity. Most network operating systems support audit trails. Securing the network also requires securing the devices. This is particularly important on the devices that interconnect large parts of the enterprise, such as the backbone or campus routers. These devices should be in one spot or in secure rooms placed strategically around the enterprise. They can be engineered to generate an alarm (raise a management event) if the cases are opened, module removed or if anything is changed.

Additional levels of security are also provided by building in rules, filters and screening into the interconnecting devices. For example, a router can be configured to only allow certain IP addresses access to a segment of the backbone.

The Server

The first line of defence for the server(s) is to have the server centre in a secure location where access, by authorised personnel only, can be supervised and administration can be performed simply. The server should be attached to an uninterruptable power supply as this protects the server by filtering the mains and provides backup power should the supply fail. This allows the server to shut itself down in a controlled manner to protect data.

The servers should be protected with the level of password security applicable to the business. Upgrades to the server software should be planned, monitored and controlled with updates done out of hours. Virus protection should also be active on PC-based servers.

To ensure privacy of the data the entire contents of the database can be encrypted using either cryptogram or advanced DES encryption mechanisms. The level of encryption is dependant on government regulations. For example, SQLBase supports a minimum of 40-bit encryption.

Some database systems, *e.g.*, Oracle, can validate database users without database passwords by using information in the hosts operating system authentication mechanism. This simplifies the database security administration as it is centralised at a system level. However, the problem with this is that often the database user does not have to start a host session on the database server before using the database across the network, effectively bypassing the operating system security mechanisms.

Many client/server database systems do not have adequate password management facilities similar to those found in operating systems. This means that there is no easy way for the database administrator to ensure that users have chosen good passwords and will change them frequently. Most database systems allow users to change their passwords using simple SQL utilities. Unfortunately these SQL utilities do not force the user to verify the current password prior to changing it. Therefore, it is easy to change another user's password. Third party database password management utilities are starting to appear which address the above deficiencies.

To guard against "trial and error" login attacks to a database, servers can impose increasingly long delays in responding to a user, with additional mechanisms to alert other users in the event of a protracted attack.

Server database software vendors are now extending their products to include government specifications for multilevel data security. Database management systems can offer their own security, for example, Sybase uses stored procedures to define, limit and authorise user access. By storing the security within the database, no data can be accessed without going through the security checks.

The integrity of the database can be maintained by guaranteeing that data cannot be changed except through the control of the database. Page checksums enforce this.

The Users

The first line of defence against illegal entry to a multi-user client/server system is user identification and authentication. It follows, that the easiest way to gain illegal entry to the system is by obtaining a valid users ID and password. The problem with keeping passwords secret has been around since passwords were invented. For example, they can be discovered when:

- The user picks a short password or one that is easy to guess, such a spouse's name.
- The user keeps a list of passwords taped on the screen or in a desk drawer.
- The users share their passwords with other users.
- An attacker phones the user, posing as one of the companies IT staff, and requests the user's password to fix an unnamed problem.

To overcome this a good security policy and strong password management must be implemented. A security policy will set guidelines for minimum password length, types of passwords that can be chosen, how often passwords should be changed, and so on. Password management utilities are available to check for guessable passwords, for minimum lengths and regularly ask users to change their passwords.

There is no one solution that addresses all security issues raised when implementing a client/server system. The view to the users must be that of a single homogeneous system when the reality is that it is a system made up of multiple levels and parts, each with its own security issues.

The enterprise security policy should encompass the physical security of devices, including the physical media in which the data is carried, the backup of critical data, which should not be stored on a desktop where it can be copied to a diskette and the integrity of the network and the database through authentication and access control.

Users are often not receptive to mandatory security procedures because the very measures used to keep attackers out are just as effective at keeping the users out. One solution is to carry out as many of the security measures as possible at the network level rather than at individual workstations. This can be achieved in a network operating system that supports OSF's DCE (*i.e.*, Kerberos).

SUMMARY

- Client-server computing is a distributed computing model in which client applications request services from server processes.
- The client-server model is an extension of the object based (or modular) programming model, where large pieces of software are structured into smaller components that have well defined interfaces.
- Client/server is an architecture in which a system's functionality and its processing are divided between the client PC and a database server.
- The two-tier architecture is easier to implement and is used in small environments.
- In the three-tier architecture, the intermediate process is used for decoupling clients and servers.
- Authentication of clients is handled by using cryptographic techniques such as public key encryption or special authentication servers.
- IDL serves for declaring remotely accessible server objects in a platform-and programming language-neutral manner, but not for implementing those objects.
- CORBA objects are implemented in widely used languages such as Cþþ, C, Java, and Smalltalk.
- The three-tier client/server architecture is an evolution of the traditional two-tier model, and is receiving increased interest, particularly for large business applications.
- The most basic type of middle layer is the transaction processing monitor or TP monitor.
- An ActiveX component is a remote accessible object that has a well defined interface and is self-contained.
- A thin client is a network computer without a hard disk drive.
- A thick client is one that will perform the bulk of the processing in client/server applications.

REVIEW QUESTIONS

1. Compare client server architecture with traditional host system.
2. Explain 2-Tier client server architecture. What are its characteristics?
3. How N-Tier client-server architecture differs from 3-Tier architecture? Discuss advantages and disadvantages of both.
4. Differentiate between thick and thin client server computing.
5. What are the benefits of a client server model?

CHAPTER 9

Web Mining

Web mining is a rapid growing research area. It consists of Web usage mining, Web structure mining, and Web content mining. Web usage mining refers to the discovery of user access patterns from Web usage logs. Web structure mining tries to discover useful knowledge from the structure of hyperlinks. Web content mining aims to extract/mine useful information or knowledge from web page contents. This chapter focuses on Web Content Mining.

Web content mining is related but different from data mining and text mining. It is related to data mining because many data mining techniques can be applied in Web content mining. It is related to text mining because much of the web contents are texts. However, it is also quite different from data mining because Web data are mainly semi-structured and/or unstructured, while data mining deals primarily with structured data. Web content mining is also different from text mining because of the semi-structure nature of the Web, while text mining focuses on unstructured texts. Web content mining thus requires creative applications of data mining and/or text mining techniques and also its own unique approaches. In the past few years, there was a rapid expansion of activities in the Web content mining area. This is not surprising because of the phenomenal growth of the Web contents and significant economic benefit of such mining. However, due to the heterogeneity and the lack of structure of Web data, automated discovery of targeted or unexpected knowledge information still present many challenging research problems. In this chapter, we will examine the following important Web content mining problems and discuss existing techniques for solving these problems. Some other emerging problems will also be surveyed.

- **Data information extraction:** Our focus will be on extraction of structured data from Web pages, such as products and search results. Extracting such data allows one to provide services.
- **Web information integration and schema matching:** Although the Web contains a huge amount of data, each web site (or even page) represents similar information differently. How to identify or match semantically similar data is a very important problem with many practical applications.
- **Opinion extraction from online sources:** There are many online opinion sources, *e.g.*, customer reviews of products, forums, blogs and chat rooms. Mining opinions (especially consumer opinions) is of great importance for marketing intelligence and product benchmarking.
- **Knowledge synthesis:** Concept hierarchies or ontology are useful in many applications. However, generating them manually is very time consuming. A few existing methods that explores the information redundancy of the Web are presented. The main application is to

synthesize and organize the pieces of information on the Web to give the user a coherent picture of the topic domain.

• **Segmenting Web pages and detecting noise:** In many Web applications, one only wants the main content of the Web page without advertisements, navigation links, copyright notices. Automatically segmenting Web page to extract the main content of the pages is interesting problem. A number of interesting techniques have been proposed in the past few years.

Web mining is commonly divided into the following three sub-areas:

(*a*) Web content mining

(*b*) Web usage mining

(*c*) Web structure mining

Web content mining focuses on the discovery/retrieval of the useful information from the Web contents/data/documents, while the Web structure mining emphasizes to the discovery of how to model the underlying link structures of the Web. The distinction between these two categories isn't very clear sometimes. Web usage mining is relatively independent, but not an isolated category, which mainly describes the techniques that discover the user's usage pattern and try to predict the user's behaviours.

9.1 INTRODUCTION

Web mining is the term of applying data mining techniques to automatically discover and extract useful information from the World Wide Web documents and services. Although Web mining puts down the roots deeply in data mining, it is not equivalent to data mining. The unstructured feature of Web data triggers more complexity of Web mining. Web mining research is actually a converging area from several research communities, such as Database, Information Retrieval, Artificial Intelligence, and also psychology and statistics as well. The way to decompose Web mining into the following subtasks:

(*a*) **Resource Discovery:** the task of retrieving the intended information from Web.

(*b*) **Information Extraction:** automatically selecting and pre-processing specific information from the retrieved Web resources.

(*c*) **Generalization:** automatically discovers general patters at the both individual Web sites and across multiple sites.

(*d*) **Analysis:** analyzing the mined pattern.

In brief, Web mining is a technique to discover and analyze the useful information from the Web data. Web involves three types of data:

(*a*) data on the Web (content).

(*b*) Web log data (usage).

(*c*) Web structure data.

The data type are classified as content data, structure data, usage data, and user profile data. The figure below depicts the taxonomy for web mining.

With the recent explosive growth of the amount of content on the Internet, it has become increasingly difficult for users to find and utilize information and for content providers to classify and catalog documents. Traditional web search engines often return hundreds or thousands of results for a search, which is time consuming for users to browse. On-line libraries, search engines, and other large document repositories (*e.g.*, customer support databases, product specification databases, press release archives, news story archives, etc.) are growing so rapidly that it is difficult and costly to categorize every document manually. In order to deal with these problems, researchers look toward automated methods of working with web documents so that they can be more easily browsed, organized, and cataloged with minimal human intervention.

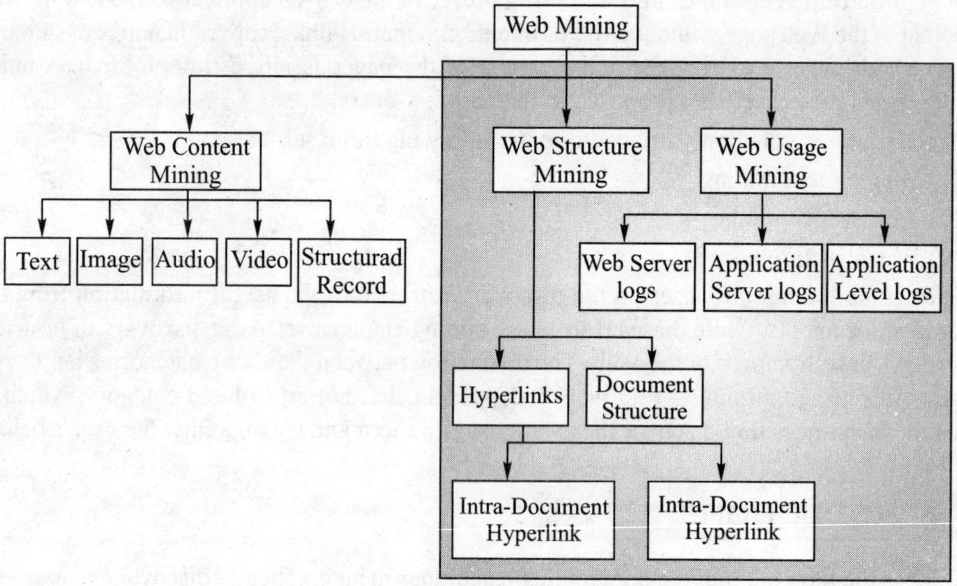

Figure 9.1. Web Mining Taxonomy

In contrast to the highly structured tabular data upon which most machine learning methods are expected to operate, web and text documents are semi-structured. Web documents have well-defined structures such as letters, words, sentences, paragraphs, sections, punctuation marks, HTML tags, and so forth. As words make up sentences, sentences make up paragraphs, and so on, but many of the rules governing the order in which the various elements are allowed to appear are vague or ill-defined and can vary dramatically between documents. It is estimated that as much as 80-85% of all digital business information, most of it web-related, is stored in non-structured formats (*i.e.*, non-tabular formats, such as those that are used in databases and spreadsheets). Developing improved methods of performing machine learning techniques on this vast amount of non-tabular, semi-structured web data is therefore highly desirable.

Clustering and classification have been useful and active areas of machine learning research that promise to help us cope with the problem of information overload on the Internet. With clustering the goal is to separate a given group of data items (the data set) into groups called clusters such that items in the same cluster are similar to each other and dissimilar to the items in other clusters. In clustering methods no labeled examples are provided in advance for training (this is called unsupervised learning). Under classification we attempt to assign a data item to a predefined category based on a model that is created from pre-classified training data (supervised learning). In more general terms, both clustering and classification come under the area of knowledge discovery in databases or data mining. Applying data mining techniques to web page content is referred to as web content mining which is a new sub-area of web mining, partially built upon the established field of information retrieval.

When representing text and web document content for clustering and classification, a vector-space model is typically used. In this model, each possible term that can appear in a document becomes a feature dimension. The value assigned to each dimension of a document may indicate the number of times the corresponding term appears on it or it may be a weight that takes into account other frequency information, such as the number of documents upon which the terms appear. This model is simple and allows the use of traditional machine learning methods that deal with numerical feature vectors in a Euclidean feature space. However, it discards information such as the order in which the terms appear, how close the terms are to each other, and so forth. By keeping this kind of structural information we could possibly improve the performance of various machine learning algorithms.

The problem is that traditional data mining methods are often restricted to working on purely numeric feature vectors due to the need to compute distances between data items or to calculate some representative of a cluster of items (*i.e.*, a centroid or center of a cluster), both of which are easily accomplished in a Euclidean space. Thus either the original data needs to be converted to a vector of numeric values by discarding possibly useful structural information or we need to develop new, customized methodologies for the specific representation.

Graphs are important and effective mathematical constructs for modeling relationships and structural information. Graphs and their more restrictive form, trees are used in many different problems, including sorting, compression, traffic flow analysis, resource allocation, etc. In addition to problems where the graph itself is processed by some algorithm (*e.g.*, sorting by the depth first method or finding the minimum spanning tree) it would be extremely desirable in many applications, including those related to machine learning, to model data as graphs since these graphs can retain more information than sets or vectors of simple atomic features. Thus much research has been performed in the area of graph similarity in order to exploit the additional information allowed by graph representations by introducing mathematical frameworks for dealing with graphs. Some application domains where graph similarity techniques have been applied include face and fingerprint recognition as well as anomaly detection in communication networks.

9.2 WEB MINING FEATURES AND FUNCTIONS

Various features of web mining and its functionalities are listed below:

- Content mining extracts patterns from online information, such as HTML files, images, or e-mails.
- Web structure mining focuses on using the analysis of the link structure of the web to identify preferable documents.
- Web usage mining for user interactions whenever requests for resources are received.
- Concise notation based on familiar ANSI standard SQL, including joins, grouping, sorting, and set operations.
- Full featured IDE includes syntax highlighting, graphical execution, real time result delivery, and network monitoring.
- Extraction techniques for unstructured, semi-structured, and structured data.
- Reads and writes common file formats such as HTML, XML, PDF, DOC, CSV, TSV, images, databases, etc.
- Integration options include Java, .NET, ActiveX, C++, and allowing queries to become Web services.
- Facilities for error trapping and reporting.
- XML, including extraction of data using XPath and transformation of input or output using XSLT.

- Transparent support for Web functionality such as scripts, forms, cookies, user agents, frames, tables, authentication, etc.
- Development environment for organizations to create their own Web mining solutions.
- Parallel deployment engine extracts information at high speed.

9.3 WEB CONTENT MINING

Web content mining describes the automatic search of information resource available online, and involves mining web data contents. In the Web mining domain, Web content mining essentially is an analog of data mining techniques for relational databases, since it is possible to find similar types of knowledge from the unstructured data residing in Web documents.

The Web document usually contains several types of data, such as text, image, audio, video, metadata and hyperlinks. Some of them are semi-structured such as HTML documents or a more structured data like the data in the tables or database generated HTML pages, but most of the data is unstructured text data. The unstructured characteristics of Web data force the Web content mining towards a more complicated approach. The Web content mining is differentiated from two different points of view: Information Retrieval View and Database View.

Multimedia data mining is part of the content mining, which is engaged to mine the high-level information and knowledge from large online multimedia sources. Multimedia data mining on the Web has gained many researchers' attention recently. Working towards a unifying framework for representation, problem solving, and learning from multimedia is really a challenge, this research area is still in its infancy indeed, many works are waiting to be done.

9.4 WEB STRUCTURE MINING

Most of the Web information retrieval tools only use the textual information, while ignore the link information that could be very valuable. The goal of Web structure mining is to generate structural summary about the Web site and Web page. Technically, Web content mining mainly focuses on the structure of inner-document, while Web structure mining tries to discover the link structure of the hyperlinks at the inter-document level. Based on the topology of the hyperlinks, Web structure mining will categorize the Web pages and generate the information, such as the similarity and relationship between different Web sites.

Web structure mining can also have another direction – discovering the structure of Web document itself. This type of structure mining can be used to reveal the structure (schema) of Web pages; this would be good for navigation purpose and make it possible to compare/integrate Web page schemes. This type of structure mining will facilitate introducing database techniques for accessing information in Web pages by providing a reference schema.

The structural information generated from the Web structure mining includes the follows:

- the information measuring the frequency of the local links in the Web tuples in a Web table;
- the information measuring the frequency of Web tuples in a Web table containing links that are interior and the links that are within the same document;
- the information measuring the frequency of Web tuples in a Web table that contains links that are global and the links that span different Web sites;
- the information measuring the frequency of identical Web tuples that appear in a Web table or among the Web tables.

In general, if a Web page is linked to another Web page directly, or the Web pages are neighbors, we would like to discover the relationships among those Web pages. The relations may fall in one of the types, such as they are related by synonyms or ontology, they may have similar contents, both of them may sit in the same Web server therefore created by the same person. Another task of Web structure mining is to discover the nature of the hierarchy or network of hyperlinks in the Web sites of a particular domain. This may help to generalize the flow of information in Web sites that may represent some particular domain, therefore the query processing will be easier and more efficient.

Web structure mining has a natural relation with the Web content mining, since it is very likely that the Web documents contain links, and they both use the real or primary data on the Web. It's quite often to combine these two mining tasks in an application.

The structure of a typical web graph consists of web pages as nodes, and hyperlinks as edges connecting related pages. Web structure mining is the process of discovering structure information from the web. This can be further divided into two kinds based on the kind of structure information used.

Hyperlinks

A hyperlink is a structural unit that connects a location in a web page to a different location, either within the same web page or on a different web page. A hyperlink that connects to a different part of the same page is called an *intra-document hyperlink*, and a hyperlink that connects two different pages is called an *inter-document hyperlink.*

Document Structure

In addition, the content within a Web page can also be organized in a tree structured format, based on the various HTML and XML tags within the page. Mining efforts have focused on automatically extracting document object model (DOM) structures out of documents.

9.5 WEB USAGE MINING

Web usage mining tries to discover the useful information from the secondary data derived from the interactions of the users while surfing on the Web. It focuses on the techniques that could predict user behavior while the user interacts with Web. The potential strategic aim in each domain is divided into mining goal as:

- prediction of the user's behavior within the site,
- comparison between expected and actual Web site usage,
- adjustment of the Web site to the interests of its users.

There are no definite distinctions between the Web usage mining and other two categories. In the process of data preparation of Web usage mining, the Web content and Web site topology will be used as the information sources, which interacts Web usage mining with the Web content mining and Web structure mining. Moreover, the clustering in the process of pattern discovery is a bridge to Web content and structure mining from usage mining.

Web usage mining is the application of data mining techniques to discover usage patterns from Web data, in order to understand and better serve the needs of Web-based applications. Web usage mining is parsed into three distinctive phases:

- preprocessing
- pattern discovery
- pattern analysis.

This is an excellent approach to define the usage mining procedure. It also clarifies the research sub direction of the Web usage mining, which facilitates the researchers to focus on each individual process with different applications and techniques. Figure 9.2 depicts the architecture of web usage mining. The architecture of the Web Usage Mining encompasses these three-phase processing.

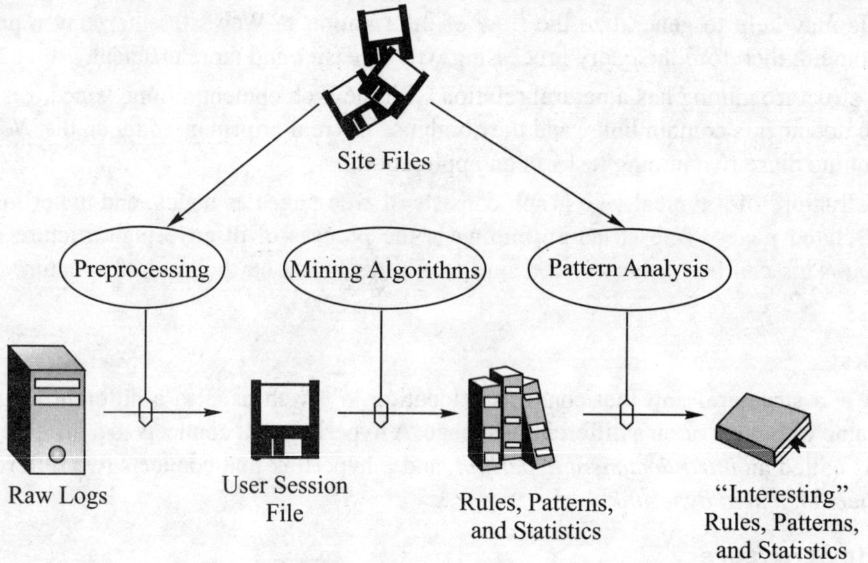

Figure 9.2. Web Usage Mining Architecture

A. Data Preprocessing for Mining

From the technique point of view, Web usage mining is the application of data mining techniques to usage logs (secondary Web data) of large Web data repositories. The purpose of it is to produce results that can be used in the design tasks such as Web site design, Web server design and of navigating through a Web site. However, before applying the data mining algorithm, we must perform a data preparation to convert the raw data into the data abstraction necessary for the further process. The data can be collected at the server-side, client-side, proxy servers, or obtained from database. For each type of data collection, the difference is not only the location, but also the available data type, the segment of population from which the data was collected and the method of implementation. The information sources available to mine include Web usage logs, Web page descriptions, Web site topology, user registries, and questionnaire. It's natural to think that the preprocess has three different conversions: Usage converting, Content converting, and Structure converting.

Since the data abstraction is very important in the data preprocess, it's necessary to clarify the definitions of the related data abstractions before the description of the different type of the data converting. The following definitions are necessary for the clarification of web usage mining.

User: The principal using a client to interactively retrieve and render resources or resource manifestations.

Page view: Visual rendering of a Web page in a specific client environment at a specific point in time.

Click stream: A sequential series of page view request.

User session: A delimited set of user clicks (click stream) across one or more Web servers.

Server session (visit): A collection of user clicks to a single Web server during a user session; also called a visit.

Episode: A subset of related user clicks that occur within a user session.

Content Preprocessing

Content preprocessing is the process of converting text, image, scripts and other files into the forms that can be used by the usage mining. It's not hard to understand that the Web content can be used to filter the input to, or output from the pattern discovery algorithm. For the content of static page views, the preprocessing can be easily done by parsing the HTML and reformatting the information or running additional algorithm as desired. It would be much more complicated to the content of dynamic page views. To perform the preprocessing, the content of each page view must be "assembled", either by an HTTP request from a crawler, or a combination of template, script, and the database accesses.

Structure Preprocessing

The structure of a Web site is formed by the hyperlinks between page views. The structure preprocessing can be treated similar as the content preprocessing. However, each server session may have to construct a different site structure than others.

Usage Preprocessing

The inputs of the preprocessing phase may include the Web server logs, referral logs, registration files, index server logs, and optionally usage statistics from a previous analysis. The outputs are the user session file, transaction file, site topology, and page classifications. It's always necessary to adopt a *data cleaning* techniques to eliminate the impact of the irrelevant items to the analysis result. The usage preprocessing probably is the most difficult task in the Web Usage Mining processing due to the incompleteness of the available data. Without sufficient data, it is very difficult to *identify the users*. The easiest way to improve the data quality is to get user cooperation, but it's not easy at all. There exists a conflict between the analysis needs of the analysts (who want more detailed usage data collected), and the privacy needs of the individual users (who want as little data collected as possible).

However, the heuristics and statistics methods can be used to improve the quality of the Web usage data. We may find some approach to solve the problem, but it is impossible to avoid the misidentification completely, since the web is so dynamic and versatile. For example, any page view accessed through a client or proxy- level cache will not be "visible" from the server side, and the only verifiable method of tracking ached page views is to monitor usage from the client side.

The *session identification* is also a part of the usage preprocessing. The goal of it is to divide the page accesses of each user, who is likely to visit the Web site more than once, into individual sessions. The simplest way to do is to use a timeout to break a user's click-stream into session. The thirty minutes is used as a default timeout by many commercial products. Another problem is named as *path completion*, which indicates the determining if there are any important accesses missed in the access log. The methods used for the user identification can be used for path completion. The final procedure of the preprocessing is *formatting*, which is a preparation module to properly format the sessions or transactions.

B. Pattern Discovery

This is the key component of the Web mining. Pattern discovery converges the algorithms and techniques from several research areas, such as data mining, machine learning, statistics, and pattern recognition. According to the techniques adopted in this area, this process is discussed in the separate subsections as follows.

Statistical Analysis

Statistical techniques are the most powerful tools in extracting knowledge about visitors to a Web site. The analysts may perform different kinds of descriptive statistical analyses based on different variables when analyzing the session file. By analyzing the statistical information contained in the periodic Web system report, the extracted report can be potentially useful for improving the system performance, enhancing the security of the system, facilitation the site modification task, and providing support for marketing decisions.

Association Rules

In the Web domain, the pages, which are most often referenced together, can be put in one single server session by applying the association rule generation. Association rule mining techniques can be used to discover unordered correlation between items found in a database of transactions. The association rules refer to sets of pages that are accessed together with a support value exceeding some specified threshold. The support is the percentage of the transactions that contain a given pattern. The Web designers can restructure their Web sites efficiently with the help of the presence or absence of the association rules. When loading a page from a remote site, association rules can be used as a trigger for prefetching documents to reduce user perceived latency.

Clustering

Clustering analysis is a technique to group together users or data items (pages) with the similar characteristics. Clustering of user information or pages can facilitate the development and execution of future marketing strategies. Clustering of users will help to discover the group of users, who have similar navigation pattern. It's very useful for inferring user demographics to perform market segmentation in E-commerce applications or provide personalized Web content to the individual users. The clustering of pages is useful for Internet search engines and Web service providers, since it can be used to discover the groups of pages having related content.

Classification

Classification is the technique to map a data item into one of several predefined classes. In the Web domain, Web master or marketer will have to use this technique if he/she want to establish a profile of users belonging to a particular class or category. This requires extraction and selection of features that best describe the properties of a given class or category. The classification can be done by using supervised inductive learning algorithms such as decision tree classifiers, naïve Bayesian classifiers, *k*-nearest neighbour classifier, Support Vector Machines etc.

Sequential Pattern

This technique intends to find the inter-session pattern, such that a set of the items follows the presence of another in a time-ordered set of sessions or episodes. It's very meaningful for the Web marketer to predict the future trend, which help to place advertisements aimed at certain user groups. Sequential patterns also include some other types of temporal analysis such as trend analysis, change point detection, or similarity analysis.

Dependency Modelling

The goal of this technique is to establish a model that is able to represent significant dependencies among the various variables in the Web domain. The modelling technique provides a theoretical framework for analyzing the behavior of users, and is potentially useful for predicting future Web resource consumption.

C. Pattern Analysis

Pattern Analysis is a final stage of the whole Web usage mining. The goal of this process is to eliminate the irrelative rules or patterns and to extract the interesting rules or patterns from the output of the pattern discovery process. The output of Web mining algorithms is often not in the form suitable for direct human consumption, and thus need to be transform to a format can be assimilate easily. This can be done with the help of some analysis methodologies and tools. There are two most common approaches for the patter analysis. One is to use the knowledge query mechanism such as SQL, while another is to construct multi-dimensional data cube before perform OLAP operations. All these methods assume the output of the previous phase has been structured. There are more techniques coming out in recent years, such as visualization etc. Although there are quite a few commercial analysis applications available and many more are free on the Web, most of them are dislike by users, considered too slow, inflexible, difficult to maintain and limited in the functionality. To develop the efficient, flexible, and powerful tools, lots of work needs to be done for both researcher and developer.

9.6 PERSONALIZATION VS. USER NAVIGATION PATTERN

The applications of Web usage mining can be classified into two main streams: personalized vs. impersonalized. Personalized means learning a user profile of user modeling in adaptive interfaces, while impersonalized means learning user navigation pattern. With the technique of personalization, the Web user would prefer an intelligent Web server which is capable to learn their information needs and preferences. On the other hand, with the technique of learning user navigation patterns, the information providers would be glad to view the improvement of the effectiveness on their Web sites, which results in adapting the Web site design or by biasing the user's behavior towards satisfying the goals of the site.

A. Personalization

The Web provides a direct communication medium between the vendors of products and services, and their customer with very low cost. There come tremendous opportunities for ecommerce development. The Web personalization is a very important, if not necessary, part of the e-commerce. Even outside of the e-commerce, Web personalization has many applications. In the context of Web mining, personalization is the provision to the individual of tailored products, services, information or information relating to products or service. The goal of personalization systems is to provide users with what they need or want without explicit indication. Three of the major categories of existing personalization systems are

- manual decision rule systems
- collaborative filtering system
- content-based filtering system.

There are several principal elements consisting of Web personalization in the framework. They are

- the modelling of Web objects (products, service, pages etc.) and subjects (users),
- categorization of objects and subjects,
- mapping between and across objects and/or subjects,
- determination of the set of actions to be recommended for personalization.

The overall process of usage-based personalization is divided into two components: offline component vs. online component. The offline component is consisted of the data preparation and specific usage mining tasks. Online component uses the discovered patterns to provide personalized content to users, based on their current navigational activity. Currently, the system relies on only anonymous usage data provided by Web server logs and hypertext structure of a site, and provides a list of recommended hypertext links to a user while browsing through a Web site.

B. User Navigation Pattern

The research of user navigation pattern focuses on the techniques to study the user behavior when navigating within a web site. While the World Wide Web turns to be the largest information resource available online, awareness of the user navigation preferences becomes an essential step. It is not only in the process of customizing and adapting the site's interface for individuals, but also in improving the site's static structure of the underlying hypertext system as well. Good knowledge on the way of visitors navigate in a web site could prevent disorientation and help the provider to place the information properly.

Analysis of user behavior has two aspects, one concerning the interests of the users and the accessed information, the other concerning the way of accessing the information. The first aspect is solved by techniques for the construct of user profiles and is not specific to the Web usage, while the second one is address by analyzing Web server logs, which falls in the field of the Web usage mining.

9.7 PRIVACY ON THE WEB

Due to the massive growth of the e-commerce, privacy becomes a sensitive topic and attracts more and more attention recently. The basic goal of Web mining is to extract information from data set for business needs, which determines its application is highly customer-related. As mentioned in the above section, there exists unavoidable conflict between the Web user and the administrator in the view of privacy. From the administrator's point of view, many of the uses of data mining are innocent, such as the data analysis to detect hidden behavioral patterns to allow supermarkets to arrange items in the ways that will encourage customers to buy more of certain products or to look for seasonal buying variations.

However, from individual point of view, many users believe that some applications of Web mining, may raise privacy concern, such as junk mails stuck mail account or personal information divulged during online shopping. The privacy concern has become the most critical concern for the Web user, and e-commerce developer. The lack of regulations in the use and deployment of Web mining systems and the widely spread privacy abuses reports related to data mining has made privacy a hot iron like never before. Privacy touches a central nerve with people and there are no easy solutions. To solve the problem, the privacy legislation is as important as the technique efforts.

Legislation efforts

In 1995, the European Union passed its Directive on Data Protection that introduces privacy protection applying to the private sector. The Directive required member countries to adopt national data protection laws that meet the standards of the Directive within three years (by 1998). The European Union's European Data Protection Directives limits access to Internet based customer information. European companies can use data about customers to profile, but those profiles are encrypted to block out customers' names. Meanwhile, the Directive prohibits member countries from transferring personal information to a non-member country or to a business located in a non-member country, if the non-member country's laws do not provide adequate protection for personal information. Unfortunately in U.S. there is no unifying framework in place, although a Congress develop legislation has been recommended by U.S. Federal Trade Commission to regulate the personal information being collected at Web sited.

Technology Development

While there are great efforts to address privacy issues by the legislative and regulation bodies, many researchers are working on new technologies to better protect consumers' privacy. Researchers at Xerox Corp.'s Palo Alto Research Centre have created an algorithm that designed to keep the behavior of online shoppers hidden from Web site operator. Encirp, a vendor of marketing software designed to work in the electronic billing environment, uses an engine that function on the *consumers' desktop* to sidestep the privacy trap. It protects consumers' privacy by avoiding centralized data storage by the service provider and provides personalized interface through the engine and data stored at the consumers' desktops.

The main challenge is yet to come up with guidelines and rules. With the rules and guidelines, site administrator may perform various analyses on the usage data without compromising the identity of an individual user. W3C has initiated a project called Platform for Privacy Preferences (P3P), which provides a protocol try to solve the conflict between Web users and the site administrators.

9.8 WEB MINING AND WEB SECURITY

In recent years, Web mining has emerged as an important branch of data mining. This is mainly due to the tremendous amount of information available from the Web, which attracted many research communities, and the recent interest of e-commerce. Web mining has followed three main directions based on which part of the Web to mine. *Web content mining*, *Web structure mining* and *Web usage mining*. Web content mining describes the process of extracting useful information from millions of sources across the Web. This includes the content of Web documents as well as data/documents that are accessible from the Web. Web content data can be a combination of unstructured data such as free texts, and semi-structured data such as text extracted from HTML files and structured data in the tables or database generated HTML pages. By large the information from the Web is non-structured. Research conducted in Web content mining mainly led by IR and DB communities have focused on providing a higher level of organization to the data available on the Web.

Web structure mining uses the link structure of the Web for the purpose of organizing, visualizing and searching. Web usage mining exploits information derived from user interaction with the system to discover user's patterns in order to understand and better serve the needs of Web users. Web usage data mainly includes information automatically generated by the system and stored in access logs. It also includes data collected from other sources resulting from the user interaction such as proxy server logs, registration data, user profiles. Each of these Web mining categories concentrate on mining one main type of Web data but all three use the same subtasks to perform the mining process namely, *resource finding*, *preprocessing*, *pattern discovery* or *generalization* and *pattern analysis*. Resource finding refers to the process of retrieving information that is accessible from the Web resources such as data in newsletters, or text data extracted from HTML documents. Data preprocessing refers to the set of transformations applied to the original information retrieved in the previous step to get it ready or the mining process. This includes cleaning information from irrelevant data such as stop words, properly identifying the items that are relevant to the mining process; and presenting the information in a desired representation such as extracting textual information from HTML documents and transforming them into relational tables.

Using mining techniques for Web security needs to be investigated.

9.9 ADVANTAGES OF WEB MINING

Web mining is attractive for companies because of several advantages. In the most general sense it can contribute to the increase of profits, be it by actually selling more products or services, or by minimizing the costs. In order to do this, marketing intelligence is required. This intelligence can focus on marketing strategies and competitive analysis or on the relationship with the customers. The different kinds of web data that are somehow related to customers will then be categorized and clustered to build detailed customer profiles. This not only helps companies to retain current customers by being able to provide more personalized services, but it also contributes in the search for potential customers. The scenario clearly illustrates this.

Content and Structure Mining

One of the things that make the web so special is that its content and structure data largely are publicly available. So in theory everybody can perform content and structure mining on the web, provided they have the right knowledge and sufficient means for it. Content and structure mining tools are developed by organizations that specialize in web search and data mining technologies. Techniques like finding related words based on common occurrence within the same page, result in a larger number of interesting patterns. Applying data mining techniques to web content data, could therefore result in better ways of finding relevant information on the web. Structure mining can aid to this goal, by identifying popular sites (so-called 'authorities'), for example, by analyzing the number of links that refer to a particular site.

Web content and structure mining are not only used to improve the quality of public search engines. Special search services can also be offered. Content and structure mining tools can for instance track down online misuse of brands, or analyze the content and structure of competitive web sites in detail to gain some strategic advantage. With content and structure mining tools, things like online curriculum vitae or personal homepages can be collected. After interpreting the personal data found on personal pages this information could be used for marketing purposes. Profiles on potential customers can be produced and more detailed information is added to profiles of current customers. So mining the web not only contributes to acquiring new customers, it can also aid in retaining existing ones.

Usage Mining

Just like content and structure mining, usage mining also provides marketing intelligence. In contrast to content and structure data, web usage data is, however, not publicly available. Only those who provide the user access on the web and those who own the sites visited by the user are able to produce transaction logs. The advantages of usage mining therefore lie in building profiles based on this transactional data. Web logs provide an exciting new way of collecting information on visitors. In a way a site owner can actually see what the visitor is looking at. Any action that tailors the web experience to a particular user, or set of users, can be seen as 'web personalization'. Web personalization is often regarded to be an indispensable part of e-commerce. The ability to track web users' browsing behavior down to individual mouse clicks makes it possible to personalize services for individual customers on a massive scale. This 'mass customization' of services not only helps customers by satisfying their needs, but also results in customer loyalty.

Due to a more personalized and customer-centered approach, the content and structure of a web site can be evaluated and adapted to the customer's preferences and the right offers can be made to the right customer. When web sites automatically improve their organization and presentation by learning from visitor's access patterns, we speak of adaptive web sites. Web usage mining is also used to create personalized search engines, which can understand a person's search queries in a personal way by analyzing and profiling the user's search behavior. It offers better, more personalized information after filtering out the links that a user most likely will not be interested in. Thus mining web usage data can improve personalization, create selling opportunities and, ultimately, lead to more profitable relationships with customers.

Let us now have a look at the way in which individual web users can benefit from all those advantages. Clearly web users benefit from web mining when the techniques are used to improve the quality of public or personalized search engines. It assists them while navigating through the huge and ever- growing web. When companies provide more personalized services on their (adaptive) web sites, the individual web user can benefit from offers and services that are adjusted to his personal needs and preferences. Some even argue that the growing accuracy of profiles will lead to less unsolicited marketing approaches, for when a company is approaching a web user, it will most likely be for something he or she actually is interested in. Apparently a lot of the advantages of web mining are based on customer profiling. It is often more cost efficient to look at a group instead of looking at each individual, because groups are cheaper and easier to approach, for instance by placing an ad in the right magazine instead of mailing every individual member of the group.

These so-called group profiles can also be of added value to individual profiles, because of the fact that some individual characteristics only become clear after looking at the individual from a group perspective. A person who never reads, may still be interested in books because he lives in a neighborhood where everybody reads a lot and he likes giving neighbors books on their birthdays. An individual profile will not show this characteristic, while the neighborhood group profile does. Furthermore, group profiling can result in better predictive values, making it easier to point out potential customers.

Web mining can obviously be quite beneficial to both businesses and individuals. However, to make sure that this technique will be further developed in a properly thought-out way, we shall also have to consider the possible objections to it. Awareness of all the possible dangers is of great importance for a well-guided development.

9.10 CONCEPTUAL MAP OF WEB MINING

As **Patricio Galeas** explains in his web page about web mining, its scope covers mainly three areas within the field of knowledge discovery:

- **Web Structure Mining (WSM):** This specialty intends to reveal the real structure of web sites through the gathering of structure related data, and mainly about its connectivity. Typically it takes into account two types of links: static and dynamic.
- **Web Content Mining (WCM):** Its goal is gathering data and identifying patterns related to the contents of the web and the searches performed on them. There are two main strategies:
  - **Web page** mining, extracting patterns directly from the contents existing in web pages. In this case the data in use can be
    - o Free **text**
    - o **HTML** pages
    - o **XML** pages
    - o **Multimedia** elements
    - o Any other type of contents existing in the web site.
  - Search results mining, intending to identify patterns in the results od the search engines.

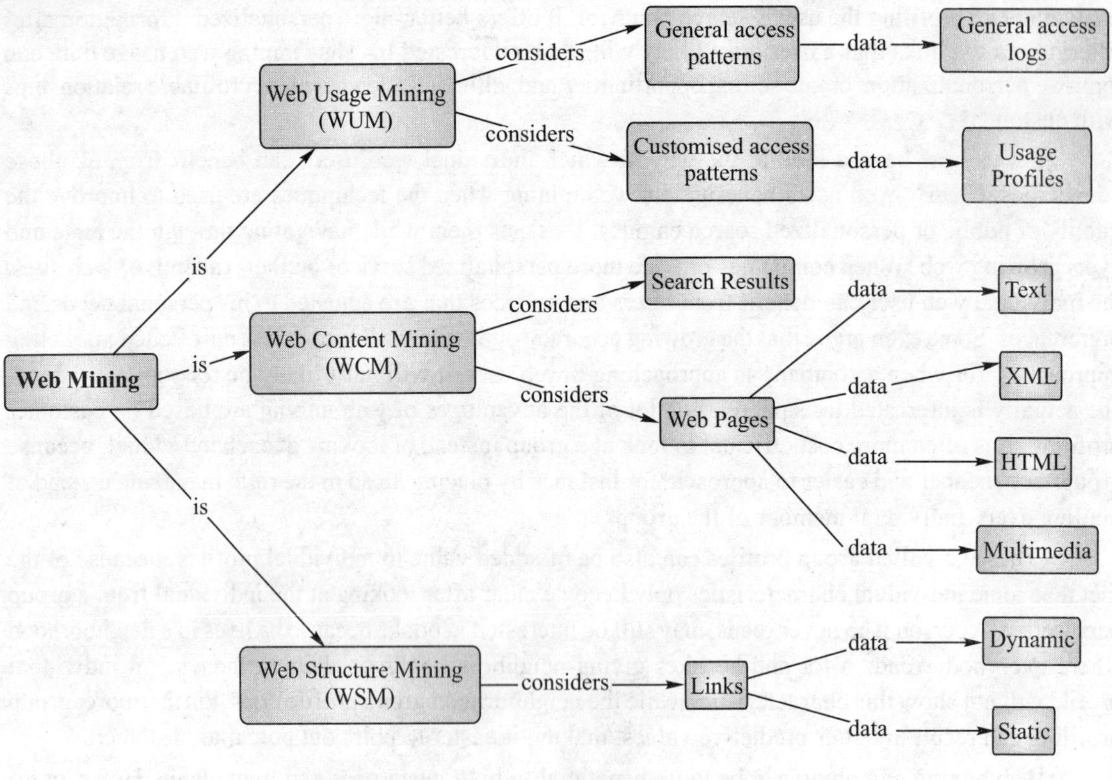

Figure 9.3. Conceptual Model of Web Mining

- **Web Usage Mining (WUM):** Here the goal is to dive into the records of the servers (log files) that store the information transactions that are performed in the web in order to find patterns revealing the usage the customers make of it. For example the most visited pages, usual visiting paths, etc. We can also distinguish here:
  - ➢ **General access pattern** tracking. Here the interest doesn't rely on the access patterns of a particular visitor but on the integration of them into trends allowing us to re-structure the web in order to facilitate our customer's access and utilization of our web site.
  - ➢ **Customized** access pattern tracking. Here what we look for is gathering data about the individual visitor's behavior and their interaction with the website. This way we can establish access/purchase profiles so that we can offer a customized experience to every customer. An archetypical case of this is amazon.com and its purchase advice and suggestions.

Web mining is a discipline with an important potential. Despite the increasing and huge volume of existing web sites the proportion of them using web mining tools to analyze their structure, contents and usage in order to improve the service to the user as the profitability of the business is still low.

9.11 WEB USAGE MINING ARCHITECTURE

The WEBMINER is a general architecture for Web usage mining. It is a system that implements parts of general architecture. The architecture divides the Web usage mining process into two main parts. The first part includes the domain dependent processes of transforming the Web data into suitable transaction form. This includes preprocessing, transaction identification, and data integration components. The second part includes the largely domain independent application of generic data mining and pattern matching techniques (such as the discovery of association rule and sequential patterns) as part of the system's data mining engine. The overall architecture for the Web mining process is depicted in Figure 9.4.

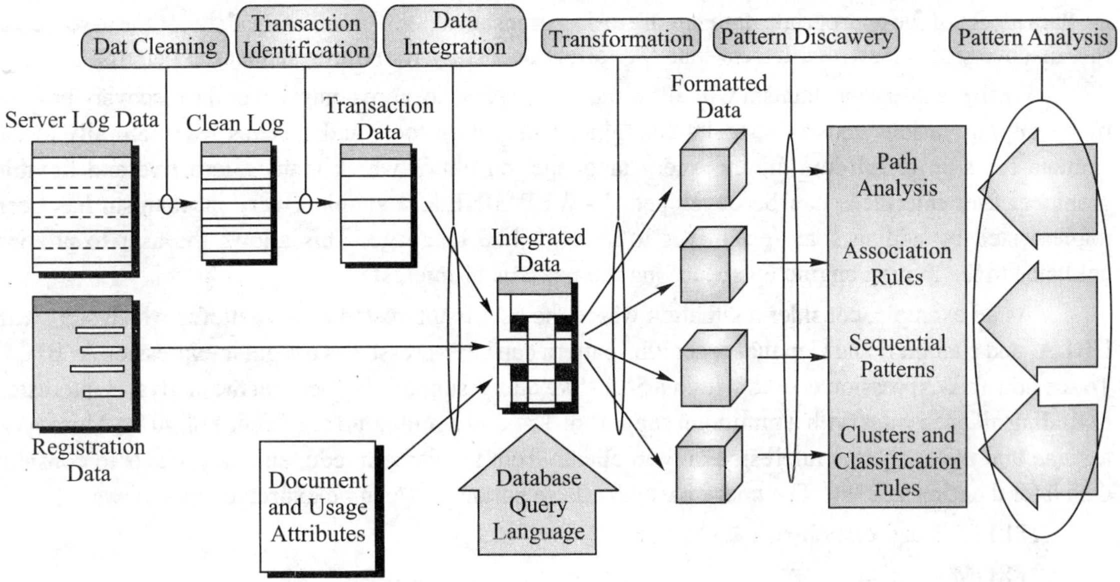

Figure 9.4. General Architecture for Web Usage Mining

Data cleaning is the first step performed in the Web usage mining process. Any of the cleaning techniques discussed previously can be used to preprocess a given Web server log. Currently, the WEBMINER system uses the simplistic method of checking filename suffixes. Some low level data integration tasks may also be performed at this stage, such as combining multiple logs, incorporating referrer logs, etc.

After the data cleaning, the log entries must be partitioned into logical clusters using one or a series of transaction identification modules. The clean server log can be thought of in two ways; either as a single transaction of many page references, or a set of many transactions each consisting of a single page reference. The goal of transaction identification is to create meaningful clusters of references for each user. Therefore, the task of identifying transactions is one of either *dividing* a large transaction into multiple smaller ones or *merging* small transactions into fewer larger ones. This process can be extended into multiple steps of *merge* or *divide* in order to create transactions appropriate for a given data mining task. A transaction identification module can be defined as either a *merge* or a *divide* module. Both types of modules take a transaction list and possibly some parameters as input, and output a transaction list that has been operated on by the function in the module in the same format as the input. The requirement that the input and output transaction format match allows any number of modules to be combined in any order, as the data analyst sees fit. The WEBMINER system currently has reference length, maximal forward reference, and time window divide modules, and a time window merge module.

Access log data may not be the only source of data for the Web mining process. User registration data, for example, is playing an increasingly important role, particularly as more security and privacy conscious client-side applications restrict server access to a variety of information, such as the client user IDs. The data collected through user registration must then be integrated with the access log data.

There are also known attributes of references pages that could be integrated into a higher level database schema. Such attributes could include page types, classification, usage frequency, page meta information, and link structures. While WEBMINER currently does not incorporate user registration data, various data integration issues are being explored in the context of Web usage mining. Once the domain-dependent data transformation phase is completed, the resulting transaction data must be formatted to conform to the data model of the appropriate data mining task. For instance, the format of the data for the association rule discovery task may be different than the format necessary for mining sequential patterns.

Finally, a query mechanism will allow the user to provide more control over the discovery process by specifying various constraints. The emerging data mining tools and systems lead naturally to the demand for a powerful data mining query language, on top of which many interactive and flexible graphical user interfaces can be developed. In WEBMINER, a simple Query mechanism has been implemented by adding some primitives to an SQL-like language. This allows the user to provide guidance to the mining engine by specifying the patterns of interest.

As an example, consider a situation where the user is interested in the patterns which start with URL A, and contain B and C in that order, this pattern can be expressed as a regular expression A*B*C*. To see how this expression is used within a SQL-like query, suppose further that the analyst is interested in finding all such rules with a minimum support of 1 % and a minimum confidence of 90 %. Moreover, assume that the analyst is interested only in clients from the domain .edu, and only wants to consider data later than Jan 1, 1996. The query based on these parameters can be expressed as follows:

```
SELECT association-rules(A*B*C*)
FROM log.data
WHERE date >= 960101 AND domain = "edu" AND
support = 1.0 AND confidence = 90.0
```

This information from the query is used to reduce the scope, and thus the cost of the mining process.

9.12 USES OF WEB MINING

This technology has enabled ecommerce to do personalized marketing, which eventually results in higher trade volumes.

- The predicting capability of the mining application can benefit the society by identifying criminal activities.
- The companies can establish better customer relationship by giving them exactly what they need.
- Companies can understand the needs of the customer better and they can react to customer needs faster.
- The companies can find, attract and retain customers, they can save on production costs by utilizing the acquired insight of customer requirements.
- They can increase profitability by target pricing based on the profiles created.
- They can even find the customer who might default to a competitor the company will try to retain the customer by providing promotional offers to the specific customer, thus reducing the risk of losing a customer.

SUMMARY

- Web usage mining refers to the discovery of user access patterns from Web usage logs.
- Web structure mining tries to discover useful knowledge from the structure of hyperlinks.
- Web content mining aims to extract/mine useful information or knowledge from web page contents.
- Web content mining requires creative applications of data mining and/or text mining techniques and also its own unique approaches.
- Web mining is the term of applying data mining techniques to automatically discover and extract useful information from the World Wide Web documents and services.
- With clustering the goal is to separate a given group of data items (the data set) into groups called clusters such that items in the same cluster are similar to each other and dissimilar to the items in other clusters.
- Under classification we attempt to assign a data item to a predefined category based on a model that is created from pre-classified training data.
- The Web document usually contains several types of data, such as text, image, audio, video, metadata and hyperlinks.
- The Web content mining is differentiated from two different points of view: Information Retrieval View and Database View.
- The structure of a typical web graph consists of web pages as nodes, and hyperlinks as edges connecting related pages.
- Web structure mining is the process of discovering structure information from the web.
- A hyperlink is a structural unit that connects a location in a web page to a different location, either within the same web page or on a different web page.
- A hyperlink that connects to a different part of the same page is called an *intra-document hyperlink*.
- A hyperlink that connects two different pages is called an *inter-document hyperlink*.
- Web usage mining is the application of data mining techniques to discover usage patterns from Web data, in order to understand and better serve the needs of Web-based applications.
- Content preprocessing is the process of converting text, image, scripts and other files into the forms that can be used by the usage mining.
- To perform the preprocessing, the content of each page view must be "assembled", either by an HTTP request from a crawler, or a combination of template, script, and the database accesses.
- The inputs of the preprocessing phase may include the Web server logs, referral logs, registration files, index server logs, and optionally usage statistics from a previous analysis.
- The outputs are the user session file, transaction file, site topology, and page classifications.
- Pattern discovery converges the algorithms and techniques from several research areas, such as data mining, machine learning, statistics, and pattern recognition.
- Analysis of user behavior has two aspects, one concerning the interests of the users and the accessed information, the other concerning the way of accessing the information.

REVIEW QUESTIONS

1. What are the functions of web mining? How is it different from data mining?
2. Explain the characteristic features of web mining?
3. Describe the three types of web mining processes.
4. How the privacy is maintained on the World Wide Web with respect to data mining?
5. Explain the concept of web security in relation with web mining.
6. Give the architectural view of data mining.
7. What are the advantages of data mining? What can be its possible future prospects?

■■■

CHAPTER

10 Security in DWH and Data Mining

Data warehouse (DWH) is a collection of integrated databases designed to support managerial decision-making and problem-solving functions. It contains both highly detailed and summarized historical data relating to various categories, subjects, or areas. All units of data are relevant to appropriate time horizons. DWH is an integral part of enterprise-wide decision support system, and does not ordinarily involve data updating. It empowers end-users to perform data access and analysis. This eliminates the need for the IS function to perform informational processing from the legacy systems for the end-users. It also gives an organization certain competitive advantages, such as: fostering a culture of information sharing; enabling employees to effectively and efficiently solve dynamic organizational problems; minimizing operating costs and maximizing revenue; attracting and maintaining market shares, and; minimizing the impact of employee turnovers.

For instance, the internal audit functions of a multi-campus institution builds a DWH to facilitate the sharing of strategic data, best audit practices, and expert insights on a variety of control topics. Auditors can access and analyze the DWH data to efficiently make well reasoned decisions (*e.g.*, recommend cost-effective solutions to various internal control problems). Mapping DWH architecture to artificial intelligence or neural applications also facilitate highly unstructured decision-making by the auditors. This results in timely completion of audit projects, improved quality of audit services, lower operating costs, and minimal impact from staff turnover. Implicit in the DWH design is the concept of *progress through sharing*.

Data warehouse is an integrated repository derived from multiple source databases. The data warehouse is created by either replicating the different source data or transforming them to new representation. This process involves reading, cleaning, aggregating and storing the data in the warehouse model. The software tools are used to access the warehouse for strategic analysis, decision-making, marketing types of applications. It can be used for inventory control of shelf stock in many departmental stores.

Medical and human genome researchers can create research data that can be either marketed or used by a wide range of users. The information and access privileges in data warehouse should mimic the constraints of source data. A recent trend is to create web-based data warehouses and multiple users can create components of the warehouse and keep an environment that is open to third party access and tools. Given the opportunity, users ask for lots of data in great detail. Since source data can be expensive, its privacy and security must be assured. The idea of adaptive querying can be used to limit access after some data has been offered to the user. Based on the user profile, the access to warehouse data can be restricted or modified.

We will focus on the following ideas that can contribute towards warehouse security.

(*a*) **Replication control:** Replication can be viewed in a slightly different manner than perceived in traditional literature. For example, an old copy can be considered a replica of the current copy of the data. A slightly out-of date data can be considered as a good substitute for some users. The basic idea is that either the warehouse keeps different replicas of the same items or creates them dynamically. The legitimate users get the most consistent and complete copy of data while casual users get a weak replica. Such replica may be enough to satisfy the user's need but do not provide information that can be used maliciously or breach privacy. We have formally defined the equivalence of replicas and this notion can be used to create replicas for different users. The replicas may be at one central site or can be distributed to proxies who may serve the users efficiently. In some cases the user may be given the weak replica and may be given an upgraded replica if willing to pay or deserves it.

(*b*) **Aggregation and Generalization:** The concept of warehouse is based on the idea of using summaries and consolidators. This implies that source data is not available in raw form. This lends to ideas that can be used for security. Some users can get aggregates only over a large number of records where as others can be given for small data instances. The granularity of aggregation can be lowered for genuine users. The generalization idea can be used to give users a high level information at first but the lower level details can be given after the security constraints are satisfied. For example, the user may be given an approximate answer initially based on some generalization over the domains of the database. Inheritance is another notion that will allow increasing capability of access for users. The users can inherit access to related data after having access to some data item.

(*c*) **Exaggeration and Misleading:** These concepts can be used to mutilate the data. A view may be available to support a particular query, but the values may be overstated in the view. For security concern, quality of views may depend on the user involved and user can be given an exaggerated view of the data. For example, instead of giving any specific sales figures, views may scale up and give only exaggerated data. In certain situations warehouse data can give some misleading information; information which may be partially incorrect or difficult to verify the correctness of the information. For example, a view of a company's annual report may contain the net profit figure including the profit from sales of properties (not the actual sales of products).

(*d*) **Anonymity:** Anonymity is to provide user and warehouse data privacy. A user does not know the source warehouse for his query and warehouse also does not know who is the user and what particular view a user is accessing (view may be constructed from many source databases for that warehouse). Note that a user must belong to the group of registered users and similarly, a user must also get data from only legitimate warehouses. In such cases, encryption is to be used to secure the connection between the users and warehouse so that no outside user (user who has not registered with the warehouse) can access the warehouse.

(*e*) **User Profile Based Security:** User profile is a representation of the preferences of any individual user. User profiles can help in authentication and determining the levels of security to access warehouse data. User profile must describe how and what has to be represented pertaining to the users information and security level authorization needs. The growth in warehouses has made relevant information access difficult in reasonable time due to the large number of sources differ in terms of context and representation. Warehouse can use data category details in determining the access control. For example, if a user would like to access an unpublished annual company report, the warehouse server may deny access to it.

The other alternative is to construct a view to reflect only projected sales and profit report. Such a construction of view may be transparent to the user. A server can use data given in the profile to decide whether the user should be given the access to associated graphical image data. The server has the option to reduce the resolution or later the quality of images before making them available to users.

10.1 DATA WAREHOUSE SECURITY

Data in the data warehouse consists of following factors:

(*a*) **Responsibility and Confidentiality:** The Data Warehouse contains confidential and sensitive data. In order to use its data, you must have proper authorization. Your authorization means that you have the authority to use the data and the responsibility to share stewardship of the data with the other users of the collection. Once authorized, you can access the data that you need to do your job. All authorized users are cautioned that they are entrusted to use the data they retrieve from the Warehouse with care. Confidential data should not be released to others except for those with a "legitimate need to know."

(*b*) **Querying Data with Security Restrictions:** If you execute a query requesting data that you are not authorized to access, you will get results which may be incomplete because they are missing the data you are not allowed to access. If your authorization is limited to a specific set of data, be sure when querying the data that your record selection conditions include your security restrictions. For example, if you are authorized to access some data for a particular department, one of your record selection conditions should state something like "If Organization = 'My Organization'," where My Organization is the code of your department. This will document why the query gets the results it does, and will also help your query run faster.

(*c*) **Data Stewardship Best Practices:** One of the most effective ways to safeguard the privacy and security of information is to make sure that it is only shared with people who clearly need it to do their jobs. This seemingly simple safeguard is sometimes though not so simple to implement. The best practices focus on procedures for granting, reviewing and terminating data user access as well as appropriate training for data users. Data stewards of all systems, and particularly large systems with sensitive information, should ensure that they are adhering to the controls found in these best practices.

10.2 DATA WAREHOUSE CONTROL

Imagine your organization has just built its data warehouse. The new data warehouse environment enables you to access corporate data in the form you want, when you want it, and where you want it to solve dynamic organizational problems, or make important decisions. You no longer feel frustrated with the inability of the Information Systems (IS) function to respond quickly to your diverse needs for information. The new environment empowers you to have the information processing world by the tail, and you are exceedingly thrilled by it all!

Suddenly, a paranoid thought creeped into your head, and you asked the classic question: *What is your organization doing to identify, classify, quantify, and protect its valuable information assets?* You posed this question to the data warehouse architects and administrators. They told you that there was nothing to worry about because the in-built security measures of your data warehouse environment cannot allow the DoD systems to attack. Somewhere along the lines, you sensed that they were not convincing.

So, you put on your hacking hat and went about the process of finding the answer to your question. As a general user, you easily managed to access some powerful user tools that were presumably restricted to unlimited access users. The tools enabled you to issue complex queries which accessed numerous data, consumed enormous resources, and slowed system response time considerably. Your trusted friend, a reformed hacker, was also able to access sensitive corporate data through the Internet without much ado. He was able to disclose your exact salary, birth date, social security number, and the date of your last performance evaluation among other things.

Your findings led you to the classic answer: *Your organization, like most, is doing little or nothing to protect its strategic information assets!* Your data warehouse administrators could not pinpoint the causes of recent system problems and security breaches until you showed them the shocking results of what you and your friend had done. It was then that they admitted that security was not a priority during the development of data warehouse. Driven by the needs to complete the data warehouse project on time and within budget, and get impatient users off their backs, they did not give security requirements any thought.

Your ecstatic excitement about the new data warehouse vanished into the thick air of security concerns over your valuable corporate data. As a diligent corporate steward, you realized that it is high time for your organization to take a *reality check*!

The security requirements of the DWH environment are not unlike those of other distributed computing systems. Thus, having an internal control mechanism to assure the confidentiality, integrity and availability of data in a distributed environment is of paramount importance. Unfortunately, most data warehouses are built with little or no consideration given to security during the development phase. Achieving proactive security requirements of DW is a seven-phase process:

1. identifying data,
2. classifying data,
3. quantifying the value of data,
4. identifying data security vulnerabilities,
5. identifying data protection measures and their costs,
6. selecting cost-effective security measures,
7. evaluating the effectiveness of security measures.

These phases are part of an enterprise-wide vulnerability assessment and management program.

Phase One–Identifying the Data

The first security task is to identify all digitally stored corporate data placed in the DWH. This is often ignored, but critical phase of meeting the security requirements of the DWH environment since it forms the foundation for subsequent phases. It entails taking a complete inventory of all the data that is available to the DWH end-users. The installed ***data monitoring*** software—an important component of the DWH—can provide an accurate information about all databases, tables, columns, rows of data, and profiles of data residing in the DW environment as well as who is using the data and how often they use the data. A manual procedure would require preparing a checklist of the same information described above. Whether the required information is gathered through an automated or a manual method, the collected information needs to be organized, documented and retained for the next phase.

Phase Two—Classifying the Data

Classifying all the data in the DWH environment is needed to satisfy security requirements for data confidentiality, integrity and availability in a prudent manner. In some cases, data classification is a legally mandated requirement. Performing this task requires the involvement of the data owners, custodians, and the end-users. Data is generally classified on the basis of criticality or sensitivity to disclosure, modification, and destruction. The sensitivity of corporate data can be classified as:

- **PUBLIC (*Least Sensitive Data*):** For data that is less sensitive than *confidential* corporate data. Data in this category is usually unclassified and subject to public disclosure by laws, common business practices, or company policies. All levels of the DWH end-users can access this data (*e.g.*, audited financial statements, admission information, phone directories, etc.).
- **CONFIDENTIAL (*Moderately Sensitive Data*):** For data that is more sensitive than *public* data, but less sensitive than *top secret* data. Data in this category is not subject to public disclosure. The principle of ***least privilege*** applies to this data classification category, and access to the data is limited to a need-to-know basis. Users can only access this data if it is needed to perform their work successfully (*e.g.*, personnel/payroll information, medical history, investments, etc.).
- **TOP SECRET (*Most Sensitive Data*):** For data that is more sensitive than *confidential* data. Data in this category is highly sensitive and mission-critical. The principle of *least privilege* also applies to this category — with access requirements much more stringent than those of the *confidential* data. Only high-level DWH users (*e.g.*, unlimited access) with proper security clearance can access this data (*e.g.*, R&D, new product lines, trade secrets, recruitment strategy, etc.). Users can access only the data needed to accomplish their critical job duties.

Regardless of which categories are used to classify data on the basis of sensitivity, the universal goal of data classification is to rank data categories by increasing degrees of sensitivity so that different protective measures can be used for different categories. Classifying data into different categories is not as easy as it seems. Certain data represents a mixture of two or more categories depending on the context used (*e.g.*, time, location, and laws in effect). Determining how to classify this kind of data is both challenging and interesting.

Phase Three–Quantifying the Value of Data

In most organizations, senior management demands to see the *smoking gun* (*e.g.*, cost-vs-benefit figures, or hard evidence of committed frauds) before committing corporate funds to support security initiatives. Cynic managers will be quick to point out that they deal with *hard reality*—not *soft variables* invented hypothetically. Quantifying the value of sensitive data warranting protective measures is as close to the smoking gun as one can get to trigger senior management's support and commitment to security initiatives in the DWH environment.

The quantification process is primarily concerned about assigning "street value" to data grouped under different sensitivity categories. By itself, data has no intrinsic value. However, the definite value of data is often measurable by the cost to:

(*a*) reconstruct lost data,

(*b*) restore the integrity of corrupted, fabricated, or intercepted data,

(*c*) not make timely decisions due to denial of service,

(*d*) pay financial liability for public disclosure of confidential data.

The data value may also include lost revenue from leakage of trade secrets to competitors, and advance use of secret financial data by rogue employees in the stock market prior to public release.

Measuring the value of sensitive data is often a Herculean task. Some organizations use simple procedures for measuring the value of data. They build a spreadsheet application utilizing both qualitative and quantitative factors to reliably estimate the *annualized loss expectancy* (ALE) of data at risk. Measuring the value of strategic information assets based on accepted classification categories can be used to show what an organization can save if the assets are properly protected, or lose (annual dollar loss) if it does not act to protect the valuable assets.

Phase Four—Identifying Data Vulnerabilities

This phase requires the identification and documentation of vulnerabilities associated with the DWH environment. Some common vulnerabilities of DWH include the following:

- ***In-built DBMS Security:*** Most data warehouses rely heavily on in-built security that is primarily VIEW-based. The VIEW-based security is inadequate for the DWH because it can be easily bypassed by a *direct dump* of data. It also does not protect data during the transmission from servers to clients—exposing the data to unauthorized access. The security feature is equally ineffective for the DWH environment where the activities of the end-users are largely unpredictable.
- ***DBMS Limitations:*** Not all database systems housing the DWH data have the capability to concurrently handle data of different sensitivity levels. Most organizations, for instance, use one DWH server to process top secret and confidential data at the same time. However, the programs handling high top security data may not prevent leaking the data to the programs handling the confidential data, and limited DWH users authorized to access only the confidential data may not be prevented from accessing the top secret data.
- ***Dual Security Engines:*** Some data warehouses combine the in-built DBMS security features with the operating system access control package to satisfy their security requirements. Using dual security engines tends to present opportunity for security lapses and intensify the complexity of security administration in the DWH environment.
- ***Inference Attacks:*** Different access privileges are granted to different DWH users. All users can access public data, but only a selected few would presumably access confidential or top secret data. Unfortunately, general users can access protected data by inference without having a direct access to the protected data. Sensitive data is typically inferred from a seemingly non-sensitive data. Carrying out direct and indirect inference attacks is a common vulnerability in the DWH environment.
- ***Availability Factor:*** Availability is a critical requirement upon which the shared access philosophy of the DWH architecture is built. However, availability requirement can conflict with or compromise the confidentiality and integrity of the DWH data if not carefully considered.
- ***Human Factors:*** Accidental and intentional acts such as errors, omissions, modifications, destruction, misuse, disclosure, sabotage, frauds, and negligence account for most of the costly losses incurred by organizations. These acts adversely affect the integrity, confidentiality, and availability of the DWH data.
- ***Insider Threats:*** The DWH users (employees) represent the greatest threat to valuable data. Disgruntled employees with legitimate access could leak secret data to competitors and publicly disclose certain confidential human resources data. Rogue employees can also profit from using strategic corporate data in the stock market before such information is released to the public. These activities cause

(*a*) strained relationships with business partners or government entities,

(*b*) loss of money from financial liabilities,

(*c*) loss of public confidence in the organization, and

(*d*) loss of competitive edge.

- ***Outsider Threats:*** Competitors and other outside parties pose similar threat to the DWH environment as unethical insiders. These outsiders engage in electronic espionage and other hacking techniques to steal, buy, or gather strategic corporate data in the DWH environment. Risks from these activities include

  (*a*) negative publicity which decimates the ability of a company to attract and retain customers or market shares, and

  (*b*) loss of continuity of DWH resources which negates user productivity.

The resultant losses tend to be higher than those of insider threats.

- ***Natural Factors*:** Fire, water, and air damages can render both the DWH servers and clients unusable. Risks and losses vary from organization to organization—depending mostly on location and contingency factors.
- ***Utility Factors:*** Interruption of electricity and communications service causes costly disruption to the DW environment. These factors have a lower probability of occurrence, but tend to result in excessive losses.

A comprehensive inventory of vulnerabilities inherent in thc DWH environment needs to be documented and organized for the next phase.

Phase Five–Identifying Protective Measures and Their Costs

Vulnerabilities identified in the previous phase should be considered in order to determine cost-effective protection for the DWH data at different sensitivity levels. Some protective measures for the DWH data include:

- ***The Human Wall:*** Employees represent the front-line of defense against security vulnerabilities in any decentralized computing environment, including DWH. Addressing employee hiring, training (security awareness), periodic background checks, transfers, and termination as part of the security requirements is helpful in creating security-conscious DWH environment. This approach effectively treats the *root causes*, rather than the *symptoms*, of security problems. Human resources management costs are easily measurable.
- ***Access Users Classification:*** Classify data warehouse users as
  1. General Access Users,
  2. Limited Access Users, and
  3. Unlimited Access Users for access control decisions.
- ***Access Controls:*** Use access controls policy based on principles of *least privilege* and *adequate data protection*. Enforce effective and efficient access control restrictions so that the end-users can access only the data or programs for which they have legitimate privileges. Corporate data must be protected to the degree consistent with its value. Users need to obtain a granulated security clearance before they are granted access to sensitive data. Also, access to the sensitive data should rely on more than one authentication mechanism. These access controls minimize damage from accidental and malicious attacks.

- ***Integrity Controls:*** Use a control mechanism to
  (*a*) prevent all users from *updating* and *deleting* historical data in the DWH,
  (*b*) restrict data *merge access* to authorized activities only,
  (*c*) immunize the DWH data from power failures, system crashes and corruption,
  (*d*) enable rapid recovery of data and operations in the event of disasters, and
  (*e*) ensure the availability of consistent, reliable and timely data to the users.

These are achieved through the OS integrity controls and well tested disaster recovery procedures.

- ***Data Encryption:*** Encrypting sensitive data in the DWH ensures that the data is accessed on an authorized basis only. This nullifies the potential value of data interception, fabrication and modification. It also inhibits unauthorized dumping and interpretation of data, and enables secure authentication of users. In short, encryption ensures the confidentiality, integrity, and availability of data in the DWH environment.
- ***Partitioning:*** Use a mechanism to partition sensitive data into separate tables so that only authorized users can access these tables based on legitimate needs. Partitioning scheme relies on a simple in-built DBMS security feature to prevent unauthorized access to sensitive data in the DWH environment. However, use of this method presents some data redundancy problems.
- ***Development Controls:*** Use quality control standards to guide the development, testing and maintenance of the DWH architecture. This approach ensures that security requirements are sufficiently addressed during and after the development phase. It also ensures that the system is highly elastic (*e.g.*, adaptable or responsive to changing security needs).

The estimated costs of each security measure should be determined and documented for the next phase. Commercial packages and in-house developed applications can help in identifying appropriate protective measures for known vulnerabilities, and quantifying their associated costs or financial impact. Measuring the costs usually involves determining the development, implementation, and maintenance costs of each security measure.

Phase Six–Selecting Cost-Effective Security Measures

All security measures involve expenses, and security expenses require justification. This phase relies on the results of previous phases to assess the economic impact of corporate data at risk, and select cost-effective security measures to safeguard the data against known vulnerabilities. Selecting cost-ffective security measures is congruent with a prudent business practice which ensures that the costs of protecting the data at risk do not exceed the maximum loss of the data.

However, the cost factor should not be the only criterion for selecting appropriate security measures in the DW environment. Compatibility, adaptability and potential impact on the DW performance should also be taken into consideration. Additionally, there are two important factors. First, the economy of mechanism principle dictates that a simple, well tested protective measure can be relied upon to control multiple vulnerabilities in the DW environment. Second, data, unlike hardware and software, has the shortest life span. Thus, the principle of adequate data protection dictates that the DW data can be protected with security measures that are effective and efficient enough for the short life span of the data.

Phase Seven–Evaluating the Effectiveness of Security Measures

A winning security strategy is to assume that all security measures are *not permanently effective*. Every time we identify and select cost-effective security measures to secure our strategic information assets against certain attacks, the attackers tend to double their efforts in identifying methods to defeat our implemented security measures. The best we can do is to prevent this from happening, make the attacks difficult to carry out, or be prepared to rebound quickly if our assets are attacked. We will not be well positioned to do any of these if we do not evaluate the effectiveness of security measures on an ongoing basis.

Evaluating the effectiveness of security measures should be conducted continuously to determine whether the measures are:

1. small, simple and straightforward,
2. carefully analyzed, tested and verified,
3. used properly and selectively so that they do not exclude legitimate accesses,
4. elastic so that they can respond effectively to changing security requirements, and
5. reasonably efficient in terms of time, memory space, and user-centric activities so that they do not adversely affect the protected computing resources.

It is equally important to ensure that the DW end-users understand and embrace the propriety of security measures through an effective security awareness program. The data warehouse administrator (DWA) with the delegated authority from senior management is responsible for ensuring the effectiveness of security measures.

10.3 ENCRYPTION REQUIREMENTS IN DWH

Encrypting sensitive data in the DW environment can be done at the table, column, or row level. Encrypting columns of a table containing sensitive data is the most common and straightforward approach used. Few examples of columns that are usually encrypted include social security numbers, salaries, birth dates, performance evaluation ratings, confidential bank information, and credit card numbers. Locating individual records in a table through a standard search command will be exceedingly difficult if any of the encrypted columns serve as keys to the table. Organizations that use social security numbers as key to database tables should seriously consider using alternative randomly generated numbers as keys before encrypting the SSN column.

Encrypting only selected rows of data is not commonly used, but can be useful in some unique cases. For instance, a single encryption algorithm can be used to encrypt the ages of *some* employees who insist on non-disclosure of their ages for privacy reasons. Multiple encryption algorithms can also be used to encrypt rows of data reflecting sensitive transactions for different campuses so that geographically distributed users of the same DW can only view/search transactions related to their respective campuses. If not carefully planned, mixing separate rows of encrypted and unencrypted data and managing multiple encryption algorithms in the same DW environment can introduce chaos, including flawed data search results.

Encrypting a table is very rarely used because it essentially renders the useless data in the DW environment. The procedures required to decrypt the encrypted keys before accessing the records in a useful format are very cumbersome and cost-prohibitive.

The encryption algorithm selected for the DW environment should be able to preserve ***field type*** and ***field length*** characteristics. It should also work cooperatively with the access and analysis software package in the DW environment. Specifically, the data decryption sequence must be executed before it reaches the software package handling the standard query. Otherwise, the package could prevent decryption of the encrypted data—rendering the data useless.

Encryption Constraints

Performing data encryption and decryption on the DW server consumes significant CPU processing cycles. This results in excessive overhead costs and degraded system performance. Also, performing decryption on the DW server before transmitting the decrypted data to the client exposes the data to unauthorized access during the transmission. These problems can be minimized if the encryption and decryption functions are effectively deployed to the workstation level with greater CPU cycles available for processing.

In addition, weak encryption algorithms can give users a false sense of security. Encrypted data in the DW must be decrypted before the standard query operations can be performed. This increases the time to process a query which can irritate the end-users and force them to be distant from encryption mechanism.

10.4 CONTROL REVIEW AND DWA

The size of historical data in the DW environment grows significantly every year, while the use of the data tends to decrease dramatically. This increases storage, processing and operating costs of the DW annually. It necessitates the periodic phasing out of least used or unused data—usually after a detailed analysis of the least and most accessed data over a long time horizon. A prudent decision has to be made as to how long historical data should be kept in the DW environment before they are phased out on people. The DWA may not meet effectively these challenges without the necessary tools like activity and data monitors, resources like funds and staffing support, and management philosophy as strategic planning and management. For these reasons, the DW Administrator should be a good strategist, an effective communicator, an astute politician, and a competent technician.

The internal control review approach of the DW environment should be primarily forward-looking (emphasizing up-front prevention) as opposed to backward-looking (emphasizing after-the-fact verification). This approach calls for the use of pre-control and concurrent control assessment techniques to look at such issues as

(*a*) data quality control

(*b*) effectiveness of security management

(*c*) economy and efficiency of DW operations

(*d*) accomplishment of operational goals or quality standards

(*e*) overall DW administration.

Effective collaboration with the internal customers and use of automated control tools are essential for conducting these control reviews competently.

10.5 INFERENCE PROBLEM AND DATA MINING

Data mining (DM) is an attempt to answer the long-standing question "what does all this data mean?". Such investigations are inherently an attempt to automate the "inference problem" in database security. Inference is basically the process of establishing relationships between data sets, the same objective as data mining. That is, given that certain attributes apply to a set of data, we "know" that certain other attributes also apply to that set of data. This is equivalent to stating that one set "implies" the other. Now, in a multi-level secure (MLS) database, we do not want Low-classified data to infer High-classified data. Data mining processes cannot be used to compromise such rules. This is because each DM process must operate at a specified level (*i.e.*, Low) and must have access to the High data in order to "discover" the rule. However, such Low-to-High rules may be "common knowledge" but unknown to the database designer. Data mining could then be used to combine Low information until the tail of the common-knowledge rule is derived. This is the process of inference. Data are put together "in a surprising way" until some common-knowledge rule, relating Low and High data, can be applied.

Fortunately, data mining can be used effectively to enforce security. The most straightforward way is to search for rules relating Low and High data. We need not be concerned with chains of inferences, merely what conjunction of attributes for a High set may be implied by Low classified attributes for that set. The security officer doing this analysis has some advantages over an attacker, since he/she has access to both the High and Low data. In most systems, there is relatively little High data, so the number of rules relating High data to Low data is much fewer than the total number of possible rules.

Data Mining and Security

Data mining is the process of posing a series of appropriate queries to extract information from large quantities of data in the database. Data mining techniques can be applied to handle problems in database security. On the other hand, data mining techniques can also be employed to cause security problems. Data mining techniques include those based on rough sets, inductive logic programming, machine learning, and neural networks, among others. Essentially one arrives at some hypothesis, which is the information extracted, from examples and patterns observed. These patterns are observed from posing a series of queries; each query may depend on the response obtained to the previous queries posed.

Data mining techniques have applications in intrusion detection and auditing databases. In the case of auditing, the data to be mined is the large quantity of audit data. One may apply data mining tools to detect abnormal patterns. For example, suppose an employee makes an excessive number of trips to a particular country and this fact is known by posing some queries. The next query to pose is whether the employee has associations with certain people from that country. If the answer is positive, then the employee's behavior is flagged.

While the example shows how data mining tools can be used to detect abnormal behavior, the next example shows how data mining tools can be applied to cause security problems. Consider a user who has the ability to apply data mining tools. This user can pose various queries and infer sensitive hypothesis. That is, the inference problem occurs via data mining. There are various ways to handle this problem. One approach is as follows. Given a database and a particular data mining tool, apply the tool to see if sensitive information can be deduced from the unclassified information legitimately obtained. If so, then there is an inference problem. There are some issues with this approach. One is that we are applying only one tool. In reality, the user may have several tools available to him. Furthermore, it is impossible to cover all ways that the inference problem could occur.

Another approach is to build an inference controller which acts during run-time. As the user applies data mining tools, the inference controller will analyze the queries posed by the user and the answers, and determines the types of responses that should be released to the user for each query. The issues involved in building such an inference controller have to be determined. Thus, data mining is an area that is growing rapidly. Not only are there several prototypes, commercial products are also appearing. One needs to take advantage of these tools to handle certain problems in security. On the other hand, these tools can also cause security problems. Therefore, appropriate measures have to be taken to detect/prevent such problems.

10.6 ROUGH SETS AND DATABASE MINING

The theory of rough sets, originated by Zdzislaw Pawlak, has been developed rapidly in the past decade, and has evolved into a technology. This emerging new technology concerns the classificatory analysis of imprecise, uncertain or incomplete information. In the past few years, it has been demonstrated that rough set theory is a very effective methodology for data analysis and discovering rules in the attribute-value based domains. It is an efficient tool for database mining in relational databases.

Database mining can be defined as the process of mining for implicit, **previously known** and potentially useful information from very large databases by efficient knowledge discovery techniques. Mining is an expensive procedure, so we should retain as much information as possible from the relational databases. For examples, functional dependencies (FD) are known constraints in database. A database miner should not try to "re-discover" them. On the other hand, extensional-functional dependencies (EFD) that are not FD do need to be discovered. Even though rough set theory is more than a methodology of rule discovering, we will focus on this aspect here.

Pawlak Information Systems and Relational Databases

A *relation* consists of maps, t : A—> V = ĖV*a*'s, where *t*(*a*) *e* V*a*, A = {*a*, *b*, ...} is a set of attributes, V*a* = attribute domains. A *relational data model* consists of

1. a collection of relations,
2. a set of constraints (*e.g.*, keys, functional dependence)
3. a set of relational operators.

Two such models are equivalent if and only if the set of relations and constraints and operator can be expressed by the other set of relations, constraints and operators using relational algebra. For any two equivalent models, their sets of all view instances is the same at all time. In 70s, Pawlak and his school have investigated logical properties of information systems, which are instances of relational databases.

An *Pawlak Information System* consist of

1. U = {*x*, *y*,..} is a set of entities of S,
2. A = {*a*, *b*, ...} is a set of attributes,
3. V*a* = attribute domain; V = Union of all V*a*'s
4. To each *x*, there is a map *x* : A—> V; *x*(*a*) *e*V*a*.

Two different *x*'s may define the same map, which is not allowed in the relational model.

An information system is called a *decision table*, if A = C U D and C ∩ D = *f*, where C is called conditional attributes, and D is called decision attributes. Let B be a non-empty subset of A. B induces an equivalence relation, denoted by IND(B), on U:

x @ *y* (mod B) iff *x*(*b*) = *y*(*b*) for " *b e* B.

The partition induced by B is called a classification of U generated by B. The ordered pair A = (U, B) is called an approximation space. Abstractly let *R* be a collection of equivalence relations on U, then the pair (U, R) is called *Pawlak knowledge base*. It is not difficult to see that the family IND(B) of all possible subsets B of A naturally induces a Pawlak knowledge base, and vice versa.

10.7 ETHICAL, LEGAL AND SOCIAL ISSUES IN MEDICINE WITH DATA MINING

There are various social, ethical and legal issues pertaining to data in data warehouses and data mining during research operations.

Two primary ethical concerns pertaining to research based on medical records are obtaining informed consent maintaining the confidentiality of data. Ideally patients should understand for what their medical records will be used for, who will have their access to their records, and how their of records will be maintained before they give explicit consent., such consent is so far not required since clinicians and researchers have taken the availability of this information as granted. Thus record linkage has usually been carried out without patient consent and qualifies for exemption from review by most ethics review boards (ERBs). Numerous surveys outside India have revealed that patients are willing to support and participate in research but first he/she want to be consulted on the use of information from their medical records. They are worried about that their data could be used for marketing and insurance purposes. They are also concerned that sensitive information could be widely used and distributed without their knowledge. These concerns have led to international efforts to enhance the protection afforded to data from medical records. The health insurance portability and accountability act directs the secretary of Health and human services to establish safeguards for the privacy of individually identifiable health information. A variety of federal legislative proposals have also been developed to address the issue. The European commission has proposed in its draft directive that explicit patient consent should be obtained before each record can be used-a rule so stringent that record based research would probably stop together.

Until recently most Indian investigators could get retrospective analysis published without an ethics review, as most international journals do not insist on such clearance, Now ERB clearance is mandatory. Indian Researchers specially in using medical data too must get their retrospective studies reviewed. The guidelines of the Indian council of Medical research (ICMR) provide a waiver of informed consent if the study is of minimal risk or conducted in an emergency. The Medical council of India's code of medical ethics (MCI) also permits such waivers if the patient identity is not revealed. However all such proposals must be clear by ERBs in a formal meeting as there are very few ERBs in the country. Such research will definitely slow down.

How do These Regulations Affect Indian Doctors and Researchers

Making it mandatory for researchers to obtain explicit consent from patients before assessing their medical records, as now proposed by European commission, would prevent clinical studies that rely on personal records, with the exception of small case series. In US, HIPPA regulations appear to inhibit medical record and database research. Current HIPPA implementation strategies increase the workload for ERBs and researchers and increase the dropout rate for proposed studies when investigators are unable to meet the requirements. Researchers also feel that public money from government agencies and charitable organizations is wasted by ERBs when innocuous retrospective studies are required to do through multiple ethical reviews. The majority of publications from Indian institutes are related to medical records review. Only a few major institutes have ERBs and most of this form of research is not subjected to ethics review.

Raw medical data are voluminous and heterogeneous. Medical data may be collected from various images, interviews, with the patient, laboratory data, and the physician's observations and interpretations.

All these components may bear upon the diagnosis, prognosis, and treatment of the patient, and cannot be ignored. The major areas of heterogeneity of medical data may be organized under these headings:

- Volume and complexity of medical data
- Physician's interpretation
- Sensitivity and specificity analysis
- Poor mathematical characterization

Volume and Complexity of Medical Data

There is a need to develop methods for efficient mining in databases of images, which are more difficult than mining in purely numerical databases. As an example, imaging techniques like MRI, PET, and collection of ECG or EEG signals, can generate gigabytes of data per day. A single cardiac SPECT procedure on one patient may contain dozens of two dimensional images. In addition, an image of the patient's organ will almost always be accompanied by other clinical information, as well as the physician's interpretation (clinical impression, diagnosis). This heterogeneity requires high capacity data storage devices and new tools to analyze such data. It is obviously very difficult for an unaided human to process gigabytes of records. Although dealing with images is relatively easier for humans because we are able to recognize patterns, grasp basic trends in data, and formulate rational decisions. The stored information becomes less useful if it is not available in an easily comprehensible format. Visualization techniques will play an increasing role in this setting, since images are the easiest for humans to comprehend, and they can provide a great deal of information in a single snapshot of the results.

Importance of Physician's Interpretation

The physician's interpretation of images, signals, or any other clinical data, is written in unstructured free-text English, that is very difficult to standardize and thus difficult to mine. Even specialists from the same discipline cannot agree on unambiguous terms to be used in describing a patient's condition. Not only do they use different names (synonyms) to describe the same disease, but they render the task even more daunting by using different grammatical constructions to describe relationships among medical entities. It has been suggested that computer translation may hold part of the solution for processing the physician's interpretation. Principles of computer translation may be summarized as follows:

- Machine translation is typically composed of the following three steps: analysis of a source language sentence; transfer from one language to another; and generation of a target language sentence.
- Natural language can be regarded as a huge set of exceptional expressions as many expressions as possible must be collected in the dictionary. It is an endless job.
- Current translation systems can analyze and translate sentences composed of less than ten words. A reason for such failure is the ambiguity. Even a human cannot understand the meaning of a long sentence at the first reading.
- Grammatical rules in machine translation can be regarded as (artificial intelligence) production rules."
- These principles, suitably customized for medical text, may be required for future medical data mining applications that depend upon the physician's free-text interpretation as part of the data mining analysis.

Sensitivity and Specificity Analysis Validation

Nearly all diagnoses and treatments in medicine are imprecise, and are subject to rates of error. The usual paradigm in medicine for measuring this error is sensitivity and specificity analysis. One should distinguish between a test and a diagnosis in medicine. A test is one of many values used to characterize the medical condition of a patient; a diagnosis is the synthesis of many tests and observations that describes a pathophysiologic process in that patient. Both tests and diagnoses are subject to sensitivity/specificity analysis. Typically, the test-results are a proposed, inexpensive new test, whereas the hypothesis is either a more expensive test, regarded as definitive, or else a complete medical workup of the patient. The accuracy of a test, on the other hand, compares how close a new test value is to a value predicted by if...then rules. To classify a test example, the rule that matches it best determines the example's class membership. An accuracy test is defined as:

Accuracy = (TP+TN)/(TP+TN+FP+FN)

Sensitivity = TP/(TP+FN)

Specificity =TN/TN+FP

Poor Mathematical Characterization of Medical Data

Another unique feature of medical data mining is that the underlying data structures of medicine are poorly characterized mathematically, as compared to many areas of the physical sciences. Physical scientists collect data which they can put into formulas, equations, and models that reasonably reflect the relationships among their data. On the other hand, the conceptual structure of medicine consists of word-descriptions and images, with very few formal constraints on the vocabulary, the composition of images, or the allowable relationships among basic concepts. The fundamental entities of medicine, such as inflammation, ischemia, or neoplasia, are just as real to a physician as entities such as mass, length, or force are to a physical scientist; but medicine has no comparable formal structure into which a data miner can organize information, such as might be modeled by clustering, regression models, or sequence analysis. In its defense, medicine must contend with hundreds of distinct anatomic locations and thousands of diseases.

Privacy and Security of Human Data

Another unique feature is privacy and security concerns. For instance, government rules set guidelines for concealment of individual patient identifiers. At stake is not only a potential breach of patient confidentiality, with the possibility of Ensuing legal action; but also erosion of the physicianpatient-relationship, in which the patient is extraordinarily candid with the physician in the expectation that such private information will never be made public. By some Guidelines, concealment of identifiers must be irreversible. A related privacy issue may apply if, for example, crucial diagnostic information were to be discovered on patient data, and a patient could be treated if one could only go back and inform the patient about the diagnosis and possible cure. In some cases, this action may not be taken. Another issue is data security in data handling, and particularly in data transfer. Before the identifiers are concealed, only authorized persons should have access to the data. Since transferring the data electronically via the Internet is insecure, the identifiers must be carefully concealed even for transfers within a single medical institution from one unit to another. On the other hand, it has been noted in documents, that there are at least two legitimate research needs for re -identification of de-identified medical data: first, there is a need to prevent accidental duplicate records on the same patient from skewing research conclusions; second, there may be a compelling need to refer to original medical records to verify the correctness or to obtain additional information on specific patients. These special requirements could be managed by appropriate regulatory agencies, but they could not be met at all if the data are completely anonymous. There are four forms of patient data identification:

(*a*) **Anonymous data** are data that were collected so that the patient-identification was removed at the time the information was collected. For example, a block of tissue may be taken from an autopsy on a patient with a certain disease, to serve as control tissue-block in the histology laboratory. The patient's identifiers are not recorded at the time of specimen collection, and thus can never be recovered.

(*b*) **Anonymized data** are data that are collected initially with the patient-identifiers, which are subsequently, irrevocably removed. That is, there can never be a possibility of returning to the patient's record and obtaining additional information. This research practice has been common in the past. However, anonymized data, as described above, could be accidentally duplicated, and could not be verified for corrections or additional data.

(*c*) **De-identified data** are data that are collected initially with the patient-identifiers, which are subsequently encoded or encrypted. The patient can be re-identified under conditions stipulated by an appropriate agency, typically an Institutional Review Board (IRB).

(*d*) **Identified data** can only be collected under significant review by the institution, federal guidelines, etc., with the patient giving written informed consent. Even for public Internet distribution, identifier-encrypted data which enter the database only once are fairly safe from Attackers.

Data from multiple institutions are only as secure as the procedures from the least-secure contributing institution. Also, data from a single institution, in which there are multiple updates of the public database over time, are also less secure from a determined attacker. There are a variety of encryption protocols suitable for such purposes:

(*a*) double-brokered encryption

(*b*) one-time-pad encryption (lookup table)

(*c*) public-private encryption

The emerging federal paradigm for using de-identified medical data for research purposes is minimal risk. That is, if one employs only data that are collected in the ordinary diagnosis and treatment of patients, and there is no change in patient management as a result of the research, including no pressure on the patient to accept or refuse certain management, and no call-back for additional data that might upset the patient or next-of-kin, then the only risk of using such data is the loss of confidentiality to the patient. This is called minimal risk data, and may be possible to use in research projects with a simple exemption from the IRB.

There was a well-publicized case of a prominent researcher at a major institution a few years ago who called a family in order to verify certain data regarding a deceased patient under study; this is not allowed under the minimal risk paradigm.

Expected Benefits

Any use of patient data, even de-identified, must be justified to the IRB as having some expected benefits. Legally and ethically one cannot perform data analysis for playful or immoral purposes. However, the Internet is the cheapest and most convenient way to distribute data, and the most accessible to the public which may have legitimate reasons for access. For example, there may be rare -disease interest groups, medical watchdog groups, or even investigators with unconventional scientific perspectives, who have reasonable claims to mine the data, but who could not mount the financial and administrative resources to mine privately-held databases.

Special Status of Medicine

Finally, medicine has a special status in science, philosophy, and daily life. The outcomes of medical care are life or death, and they apply to everybody. Medicine is a necessity, not merely an optional luxury, pleasure, or convenience. Among all the professions, medicine has the longest apprenticeship. Most medical specialists require at least eleven years of training after high school graduation, and some surgical subspecialties require up to sixteen. In the USA, medical care costs consume one-seventh of the GDP. Licensed physicians represent about 0.2% of the U.S. population; the incomes for fulltime physicians are in the top several percent; and the average physician causes seven times his/her income to be spent on services ordered. The average citizen has high expectations of medicine and its practitioners. A sick person is expected to recover. Physicians are expected to be ethical, caring, and not too greedy. Medicine is a popular subject for the popular media. Medical care is sometimes risky, but when it fails, the desire for legal revenge is intense and punitive. Medical information about the individual patient is considered highly private, and the general public is extremely fearful about disclosure. We all enjoy the benefits of medical research conducted on other patients, but we are very often reluctant to contribute or release our own information for such purposes. When medical data are published it is expected that the researchers will maintain the dignity of the individual patient, and that the results will be used for socially beneficial purposes. It has been suggested that scientific truths are fundamentally amoral; they can be used for good or evil. Yet although medicine is based upon science, there are certain tests that may not be performed, certain questions that may not be asked, and certain conclusions that may not be drawn, because of medicine's special status. There has been a vigorous public debate, for example, on whether data obtained from human experimentation should be published and used. Data from similar experiments, performed on laboratory animals, would be regarded as valid biological data without any further consideration. The special status of medicine pervades our attitudes about medical data mining, as well as our attitudes about medical diagnosis and treatment.

10.8 DATA MINING AND HOMELAND SECURITY

Datam mining has become one of the key features of many homeland security initiatives. Often used as a means for detecting fraud, assessing risk, and product retailing, data mining involves the use of data analysis tools to discover previously unknown, valid patterns and relationships in large data sets. In the context of homeland security, data mining can be a potential means to identify terrorist activities, such as money transfers and communications, and to identify and track individual terrorists themselves, such as through travel and immigration records. While data mining represents a significant advance in the type of analytical tools currently available, there are limitations to its capability. One limitation is that although data mining can help reveal patterns and relationships, it does not tell the user the value or significance of these patterns. These types of determinations must be made by the user. A second limitation is that while data mining can identify connections between behaviors and/or variables, it does not necessarily identify a causal relationship. Successful data mining still requires skilled technical and analytical specialists who can structure the analysis and interpret the output.

Data mining is becoming increasingly common in both the private and public sectors. Industries such as banking, insurance, medicine, and retailing commonly use data mining to reduce costs, enhance research, and increase sales. In the public sector, data mining applications initially were used as a means to detect fraud and waste, but have grown to also be used for purposes such as measuring and improving program performance. However, some of the homeland security data mining applications represent a significant expansion in the quantity and scope of data to be analyzed. Some efforts that have attracted a higher level of congressional interest include the Terrorism Information Awareness (TIA) project and the Computer-Assisted Passenger Prescreening System II (CAPPS II) project. Other initiatives that have been the subject of congressional interest include the Multi-State Anti-Terrorism Information

Exchange (MATRIX), the Able Danger program, the Automated Targeting System (ATS), and data collection and analysis projects being conducted by the National Security Agency (NSA).

As with other aspects of data mining, while technological capabilities are important, there are other implementation and oversight issues that can influence the success of a project's outcome. One issue is data quality, which refers to the accuracy and completeness of the data being analyzed. A second issue is the interoperability of the data mining software and databases being used by different agencies. A third issue is mission creep, or the use of data for purposes other than for which the data were originally collected. A fourth issue is privacy. Questions that may be considered include the degree to which government agencies should use and mix commercial data with government data, whether data sources are being used for purposes other than those for which they were originally designed, and possible application of the Privacy Act to these initiatives. It is anticipated that congressional oversight of data mining projects will grow as data mining efforts continue to evolve.

10.9 LIMITATIONS OF DATA MINING AS A TERRORIST DETECTION TOOL

While data mining products can be very powerful tools, they are not self-sufficient applications. To be successful, data mining requires skilled technical and analytical specialists who can structure the analysis and interpret the output that is created. Consequently, the limitations of data mining are primarily data or personnel related, rather than technology-related. Although data mining can help reveal patterns and relationships, it does not tell the user the value or significance of these patterns. These types of determinations must be made by the user. Similarly, the validity of the patterns discovered is dependent on how they compare to "real world" circumstances. For example, to assess the validity of a data mining application designed to identify potential terrorist suspects in a large pool of individuals, the user may test the model using data that includes information about known terrorists. However, while possibly re-affirming a particular profile, it does not necessarily mean that the application will identify a suspect whose behavior significantly deviates from the original model.

Another limitation of data mining is that while it can identify connections between behaviors and/or variables, it does not necessarily identify a causal relationship. For example, an application may identify that a pattern of behavior, such as the propensity to purchase airline tickets just shortly before the flight is scheduled to depart, is related to characteristics such as income, level of education, and Internet use. However, that does not necessarily indicate that the ticket purchasing behavior is caused by one or more of these variables. In fact, the individual's behavior could be affected by some additional variable(s) such as occupation (the need to make trips on short notice), family status (a sick relative needing care), or a hobby (taking advantage of last minute discounts to visit new destinations).

Beyond these specific limitations, some researchers suggest that the circumstances surrounding our knowledge of terrorism make data mining an ill-suited tool for identifying (predicting) potential terrorists before an activity occurs. Successful "predictive data mining" requires a significant number of known instances of a particular behavior in order to develop valid predictive models. For example, data mining used to predict types of consumer behavior (*i.e.*, the likelihood of someone shopping at a particular store, the potential of a credit card usage being fraudulent) may be based on as many as millions of previous instances of the same particular behavior. Moreover, such a robust data set can still lead to false positives.

Data Mining Uses

Data mining is used for a variety of purposes in both the private and public sectors. Industries such as banking, insurance, medicine, and retailing commonly use data mining to reduce costs, enhance research, and increase sales. For example, the insurance and banking industries use data mining applications to detect fraud and assist in risk assessment (*e.g.*, credit scoring). Using customer data collected over several years, companies can develop models that predict whether a customer is a good credit risk, or whether an accident claim may be fraudulent and should be investigated more closely. The medical community sometimes uses data mining to help predict the effectiveness of a procedure or medicine. Pharmaceutical firms use data mining of chemical compounds and genetic material to help guide research on new treatments for diseases. Retailers can use information collected through affinity programs (*e.g.*, shoppers' club cards, frequent flyer points, contests) to assess the effectiveness of product selection and placement decisions, coupon offers, and which products are often purchased together. Companies such as telephone service providers and music clubs can use data mining to create a "churn analysis," to assess which customers are likely to remain as subscribers and which ones are likely to switch to a competitor.

In the public sector, data mining applications were initially used as a means to detect fraud and waste, but they have grown also to be used for purposes such as measuring and improving program performance. It has been reported that data mining has helped the federal government recover millions of dollars in fraudulent Medicare payments. The Justice Department has been able to use data mining to assess crime patterns and adjust resource allotments accordingly.

10.10 DATA MINING AS A THREAT TO SECURITY

The facts listed below depict how data mining is a threat to security.

- Data mining gives us "facts" that are not obvious to human analysts of the data
- Enables inspection and analysis of huge amounts of data
- Possible threats:
  - Predict information about classified work from correlation with unclassified work (*e.g.* budgets, staffing)
  - Detect "hidden" information based on "conspicuous" lack of information
  - Mining "Open Source" data to determine predictive events (*e.g.*, Pizza deliveries to the Pentagon)
- It isn't the data we want to protect, but correlations among data items

10.11 DATA MINING TO HANDLE SECURITY PROBLEMS

Data Mining can also be used to handle security problems.

- Data mining tools can be used to examine audit data and flag abnormal behavior
- Some work in Intrusion detection
  - *e.g.*, Neural networks to detect abnormal patterns
- SRI work on IDES
- Harris Corporation work
- Tools are being examined as a means to determine abnormal patterns and also to determine the type of problem
  - Classification techniques

- Can draw heavily on Fraud detection
  - Credit cards, calling cards, etc.

10.12 DATA MINING, NATIONAL SECURITY, PRIVACY AND CIVIL LIBERTIES

There has been much interest recently on using data mining for counter-terrorism applications. For example, data mining can be used to detect unusual patterns, terrorist activities and fraudulent behavior. While all of these applications of data mining can benefit humans and save lives, there is also a negative side to this technology, since it could be a threat to the privacy of individuals.

This is because data mining tools are available on the web or otherwise and even naïve users can apply these tools to extract information from the data stored in various databases and files and consequently violate the privacy of the individuals. To carry out effective data mining and extract useful information for counter-terrorism and national security, we need to gather all kinds of information about individuals. However, this information could be a threat to the individuals' privacy and civil liberties.

Privacy is getting more attention partly because of counterterrorism and national security. There is a lot of concern about national security vs. privacy in newspapers, magazines and television talk shows. This is mainly due to the fact that people are now realizing that to handle terrorism, the government may need to collect information about individuals. This is causing a major concern with various civil liberties unions. We are beginning to realize that many of the techniques that were developed for the past two decades or so on the inference problem can now be used to handle privacy. One of the challenges to securing databases is the inference problem. Inference is the process of users posing queries and deducing unauthorized information from the legitimate responses that they receive. This problem has been discussed quite a lot over the past two decades. However, data mining makes this problem worse.

Users now have sophisticated tools that they can use to get data and deduce patterns that could be sensitive. Without these data mining tools, users would have to be fairly sophisticated in their reasoning to be able to deduce information from posing queries to the databases. That is, data mining tools make the inference problem quite dangerous. While the inference problem mainly deals with secrecy and confidentiality we are beginning to see many parallels between the inference problem and what we now call the privacy problem. These privacy issues depend on the various policies and procedures enforced. That is, technical, political, as well as social issues play a role here.

Inference is the process of posing queries and deducing unauthorized information from the legitimate responses received. For example, the names and salaries of individuals may be unclassified individually, but while taken together they are classified. This means that one could retrieve names and employee numbers, and then later retrieve the salaries and employee numbers, and make the associations between names and salaries. The problem that occurs through this inference is called the inference problem.

Data Mining, Warehousing, and Inference

First let us have a motivating example where data mining tools are applied to cause security problems. Consider a user who has the ability to apply data mining tools. This user can pose various queries and infer sensitive hypotheses. That is, the inference problem occurs via data mining. There are various ways to handle this problem. Given a database and a particular data-mining tool, one can apply the tool to see if sensitive information can be deduced from the unclassified information legitimately obtained. If so, then there is an inference problem. There are some issues with this approach. One is that we are applying only one tool. In reality, the user may have several tools available to him. Furthermore, it is impossible to cover all ways that the inference problem could occur.

Another solution to the inference problem is to build an inference controller that can detect the motives of the user and prevent the inference problem from occurring. Such an inference controller lies between the data mining tool and the data source or database, possibly managed by a database system.

Next, let us consider data warehousing and inference. The inference problem becomes an issue here also. For example, the warehouse may store average salaries. A user may have access to the average salaries in the warehouse. From these average salaries, the user may infer the individual salaries stored in the data sources from which the warehouse may be built. These individual salaries may be sensitive.

Privacy Issues

With the World Wide Web, there is now an abundance of information about individuals that one can obtain within seconds. This information could be obtained through mining or just from information retrieval. Therefore, one needs to enforce controls on databases and data mining tools. This is a very difficult problem especially with respect to data mining. In summary, one needs to develop techniques to prevent users from mining and extracting information from the data whether they are on the web or on servers. It is portrayed data mining as a technology that is critical for say analysts and other users so that they can get the right information at the right time. Furthermore, they can also extract patterns previously unknown. This is all true. However, we do not want the information to be used in an incorrect manner. For example, based on information about a person, an insurance company could deny insurance or a loan agency could deny loans. In many cases these denials may not be legitimate. Therefore, information providers have to be very careful in what they release. Also, data mining researchers have to ensure that privacy aspects are addressed.

Next, let us examine the social aspects. In most cultures, privacy of the individuals is important. However, there are certain cultures where it is impossible to ensure privacy. These could be related to political or technological issues or the fact that people have been brought up believing that privacy is not critical. There are places where people divulge their salaries without thinking twice about it, but in many countries, salaries are very private and sensitive. It is not easy to change cultures overnight, and in many cases you do not want to change them, as preserving cultures is important. So what overall effect does this have on data mining and privacy? We do not have an answer to this yet as we are only beginning to look into it. We are however beginning to realize that perhaps we do have many of the technological solutions for handling privacy. That is, many of the technologies we have proposed for information security in general and secrecy and confidentiality in particular could be applied for privacy. However we have to now focus on the social aspects.

Next, let us examine the political and legal aspects. We include policies and procedures under this:

(*a*) What sort of secrecy/privacy controls should one enforce for the web?

(*b*) Should these secrecy/privacy polices be mandated or should they be discretionary?

(*c*) What are the consequences of violating the secrecy/privacy polices?

(*d*) Who should be administering these policies as well as managing and implementing them?

(*e*) How is data mining on the web impacted?

(*f*) Can one control how data is mined on the web?

(*g*) Once we have made technological advances on data mining, can we then enforce secrecy/privacy controls on the data mining tools?

(*h*) How is information transferred between countries?

Some of the issues are related to privacy and data mining, and some others are related to just privacy in general. We have raised some interesting questions on privacy issues and data mining as well as privacy in general. As mentioned earlier, data mining is a threat to privacy. The challenge is on protecting the privacy but at the same time not losing all the great benefits of data mining.

Inference Problem and Privacy

In an earlier section we discussed the inference problem. In general when we think of the inference problem we have secrecy in mind. However, many of the concepts apply for privacy also. We can define privacy constraints such as names and salaries taken together are Private and individually they are Public. Similarly names and healthcare records taken together are Private while individually they are Public.

When Inference problem is considered to be a privacy problem, then we can use the inference controller approach to address privacy. For example, we can develop privacy controllers similar to our approach to developing inference controllers. Furthermore, we can also have different degrees of privacy. For example, names and age together could be less private while names and salaries together could be more private. Names and healthcare records together could be most private. One can then assign some probability or fuzzy value associated with the privacy of an attribute or a collection of attributes. We need to investigate further as to whether the privacy controllers could process the privacy constants during database design, update and query operations.

Lot of work has been carried out on the inference problem in the past. We need to revisit this research and see whether it is applicable for the privacy problem.

Sensitive Data Mining

As we have mentioned, the challenge is to provide solutions to enhance national security but at the same time ensure privacy. The idea here is to continue with mining but at the same time ensure privacy as much as possible. Note that one mines the data and extracts patterns and trends. The privacy constraints determine which patterns are private and to what extent. For example, suppose one could extract the names and healthcare records. If we have a privacy constraint that states that names and healthcare records are private then this information is not released to the general public. If the information is semiprivate, then it is released to those who have a need to know. Essentially the inference controller approach we have discussed is one solution to achieving some level of privacy. It could be regarded to be a type of privacy sensitive data mining. These challenges will have to be addressed when handling privacy constraints.

Note that not all approaches to privacy enhanced data mining are the same. Researchers are taking different approaches to such data mining. Some have argued that privacy enhanced data mining may be time consuming and may not be scalable. However we need to investigate this area more before we can come up with viable solutions.

Civil Liberties Vs. National Security

Civil Liberties are about protecting the rights of the individual whether it is privacy rights, human rights or civil rights. There are various civil liberties unions and laws protecting the rights of individuals. There has been much debate recently among the counter-terrorism experts and civil liberties unions and human rights lawyers about the privacy of individuals. That is, gathering information about people, mining information about people, conduction surveillance activities and examining say e-mail messages and phone conversations are all threats to privacy and civil liberties. However, what are the alternatives if we are to combat terrorism effectively? Today we do not have any effective solutions. Do we wait until privacy violations occur and then prosecute or do we wait until national security disasters occur and then gather information? What is more important? Protecting nations from terrorist attacks or protecting the privacy of individuals? This is one of the major challenges faced by technologists, sociologists and lawyers. That is, how can we have privacy but at the same time ensure the safety of nations? What should we be sacrificing and to what extent?

SUMMARY

- DWH is an integral part of enterprise-wide decision support system, and does not ordinarily involve data updating.
- Data warehouse is an integrated repository derived from multiple source databases.
- The data warehouse is created by either replicating the different source data or transforming them to new representation.
- The software tools are used to access the warehouse for strategic analysis, decision-making, marketing types of applications.
- Anonymity is to provide user and warehouse data privacy.
- User profile is a representation of the preferences of any individual user.
- User profiles can help in authentication and determining the levels of security to access warehouse data.
- User profile must describe how and what has to be represented pertaining to the users information and security level authorization needs.
- Achieving proactive security requirements of DW is a seven-phase process: identifying data, classifying data, quantifying the value of data, identifying data security vulnerabilities, identifying data protection measures and their costs, selecting cost-effective security measures, evaluating the effectiveness of security measures.
- Encrypting columns of a table containing sensitive data is the most common and straightforward approach.
- Inference is the process of posing queries and deducing unauthorized information from the legitimate responses received.

REVIEW QUESTIONS

1. How data mining is related to Inference problem?
2. How data mining finds its limitations in Homeland security?
3. With an illustrative example describe how data mining can be a boon to health care systems?
4. How rough sets are related to data mining?
5. What are the Ethical, Legal and social issues in Medicine with Data Mining?
6. Explain the Limitations of Data Mining as a Terrorist Detection Tool.

APPENDIX

A

Data Mining Applications

Data mining is becoming a pervasive technology in activities as diverse as using historical data to predict the success of a marketing campaign, looking for patterns in financial transactions to discover illegal activities or analyzing genome sequences. From this perspective, it was just a matter of time for the discipline to reach the important area of computer security.

Applications Of Data Mining In Computer Security presents a collection of research efforts on the use of data mining in computer security.

Data mining has been loosely defined as the process of extracting information from large amounts of data. In the context of security, the information we are seeking is the knowledge of whether a security breach has been experienced, and if the answer is yes, who is the perpetrator. This information could be collected in the context of discovering intrusions that aim to breach the privacy of services, data in a computer system or alternatively, in the context of discovering evidence left in a computer system as part of criminal activity.

Applications Of Data Mining In Computer Security concentrates heavily on the use of data mining in the area of intrusion detection. The reason for this is twofold. First, the volume of data dealing with both network and host activity is so large that it makes it an ideal candidate for using data mining techniques. Second, intrusion detection is an extremely critical activity. **Applications Of Data Mining In Computer Security** is designed to meet the needs of a professional audience composed of researchers and practitioners in industry and graduate level students in computer science.

The application areas for data mining includes:

- A pharmaceutical company can analyze its recent sales force activity and their results to improve targeting of high-value physicians and determine which marketing activities will have the greatest impact in the next few months. The data needs to include competitor market activity as well as information about the local health care systems. The results can be distributed to the sales force via a wide-area network that enables the representatives to review the recommendations from the perspective of the key attributes in the decision process. The ongoing, dynamic analysis of the data warehouse allows best practices from throughout the organization to be applied in specific sales situations.
- A credit card company can leverage its vast warehouse of customer transaction data to identify customers most likely to be interested in a new credit product. Using a small test mailing, the attributes of customers with an affinity for the product can be identified. Recent projects have indicated more than a 20-fold decrease in costs for targeted mailing campaigns over conventional approaches.

- A diversified transportation company with a large direct sales force can apply data mining to identify the best prospects for its services. Using data mining to analyze its own customer experience, this company can build a unique segmentation identifying the attributes of high-value prospects. Applying this segmentation to a general business database such as those provided by Dun and Bradstreet can yield a prioritized list of prospects by region.
- A large consumer package goods company can apply data mining to improve its sales process to retailers. Data from consumer panels, shipments, and competitor activity can be applied to understand the reasons for brand and store switching. Through this analysis, the manufacturer can select promotional strategies that best reach their target customer segments.
- Decision trees constructed from bank-loan histories to produce algorithms to decide whether to grant a loan.
- Patterns of traveler behavior mined to manage the sale of discounted seats on planes, rooms in hotels, etc.
- Skycat and Sloan Sky Survey: clustering sky objects by their radiation levels in different bands allowed astronomers to distinguish between galaxies, nearby stars, and many other kinds of celestial objects.
- Comparison of the genotype of people with/without a condition allowed the discovery of a set of genes that together account for many cases of diabetes. This sort of mining will become much more important as the human genome is constructed.
- A pharmaceutical company can analyze its recent sales force activity and their results to improve targeting of high-value physicians and determine which marketing activities will have the greatest impact in the next few months. The data needs to include competitor market activity as well as information about the local health care systems. The results can be distributed to the sales force via a wide-area network that enables the representatives to review the recommendations from the perspective of the key attributes in the decision process.
- The ongoing, dynamic analysis of the data warehouse allows best practices from throughout the organization to be applied in specific sales situations.
- A credit card company can leverage its vast warehouse of customer transaction data to identify customers most likely to be interested in a new credit product. Using a small test mailing, the attributes of customers with an affinity for the product can be identified. Recent projects have indicated more than a 20-fold decrease in costs for targeted mailing campaigns over conventional approaches.
- A diversified transportation company with a large direct sales force can apply data mining to identify the best prospects for its services. Using data mining to analyze its own customer experience, this company can build a unique segmentation identifying the attributes of high value prospects. Applying this segmentation to a general business database such as those provided by Dun & Bradstreet can yield a prioritized list of prospects by region.
- A large consumer package goods company can apply data mining to improve its sales process to retailers. Data from consumer panels, shipments, and competitor activity can be applied to understand the reasons for brand and store switching. Through this analysis, the manufacturer.

Applications in Library and Information Service

Most people think of libraries as the little brick building in the heart of their community or the big brick building in the center of a campus. These notions greatly oversimplify the world of libraries, however. Most large commercial organizations have dedicated in-house library operations, as do schools, nongovernmental organizations, as well as local, state, and central governments. With the increasing use of the Internet and the World Wide Web, digital libraries have proliferate, and these serve a huge variety of different user audiences, *e.g.*, people interested in health and medicine, science and technology, industry and world news, law, and business. With this expanded view of libraries, two key insights arise. First, libraries are nearly always embedded within larger institutions. Corporate libraries serve their corporations, academic libraries serve their universities, and public libraries serve the general public. Second, libraries plays an important role within their institutions as repositories and providers of information resources. In the provider role, libraries represent in microcosm the intellectual, learning, and knowledge management activities of the people who comprise the institution.

This fact provides the basis for the strategic importance of library data mining: By ascertaining what users need to know and how well those needs are served, bibliomining can reveal insights that have meaning in the context of the library's host institution. Using bibliomining, libraries can ascertain what the reconstituencies want to learn, whether they find the information they seek, and whether that information satisfies their learning and knowledge needs. In corporate libraries, which serve the knowledge needs of commercial organizations, such insights can help to develop and maintain a competitive, cutting edge workforce. In special libraries, which support the research needs of government and non-governmental organizations, these insights can influence the success of policies and programs that are informed by research. In academic organizations, accurate insights into faculty and student knowledge needs can enhance the viability of the whole institution.

APPENDIX

B

Data Mining Tools

There are many data mining tools available on the web. Many of the publicly available tools contain a lot of bugs and are hard to install. Also, it is not advisable to use too many different tools together, as each of them will need its own data format, and a lot of time and effort will be wasted doing data conversions.

Weka

General Description: Weka is open source data mining software. It does not only support machine learning algorithms, but also data preparation and meta-learners like bagging and boosting. The whole suite is written in java, so it can be run on any platform. The package has three different interfaces: a command line interface, an Explorer GUI interface (which allows you to try out different preparation, transformation and modelling algorithms on a dataset), and an Experimenter GUI interface (which allows to run different algorithms in batch and to compare the results).

Functionalities: The functionalities of Weka more or less boil down to the algorithms described in Witten and Frank's data mining book. A short overview of the Weka functionalities:

- **SVM's:** only polynomial kernels are supported. Also, support vector regression is not supported.
- Decision trees: **ID3** and **C4.5** are implemented, and **M5'**: a model tree induction algorithm for predicting numeric values (each leaf node has a regression model). **PART** is a rule-learner that makes rules by building different decision trees and each time keeping the leaf with the largest coverage.
- Memory-based methods: **kNN** and **locally weighted regression**.
- Neural Networks: only **backpropagation** with momentum is supported.
- Simpler methods: **naive Bayes** (for numeric values, a normal distribution is used, but also 'kernel density estimation' can be used to avoid assuming a normal distribution) and **linear regression** are useful simple methods. **Two-class logistic regression** is also supported. The algorithm uses a 'ridge estimator'.
- Other simple methods: **decision tables**, **1R** (make a rule based only on one attribute) and **decision stump**(one-level decision trees). Although methods this simple might seem useless, they can be combined via boosting or bagging, and form a strong classifier through combining several weak ones.
- Also meta-learning schemes are supported:

1. **Bagging**
2. **Stacking:** using a range of base classifiers and a meta classifier which classifies their output.
3. **Adaboost:** a boosting method based on Freund & Schapire's Adaboost M1 method
4. **Multi ClassClassifier:** to solve multiclass problems using two-class classifiers: one classifier is built per class (it is also possible to ask for one classifier per pair of classes).
5. **CVParameterSelection:** will try out a defined range over a set of parameters for a certain classifier using cross-validation, and use the best combination to build the final model.

- Weka also includes a package that contains clustering algorithms. It supports:
  1. **The EM algorithm:** working with numeric as well as nominal values, but assuming that all attributes are independent.
  2. **Incremental clustering:** a clustering technique that builds a tree using the category utility measure
  3. **kMeans** is also provided.
- **Association rules:** the **APriori** algorithm is supported.
- **Some data preprocessing support is provided:** You can add new attributes (based on calculations of existing ones), transform attribute values, manually select certain attributes, discretise numeric values, remove attributes with only one distinct value, select records on the basis of attribute values, transform nominal values into binary ones, merge two nominal values into one, normalise numeric values, randomise the order of the dataset, replace missing values by the mean or the mode, create random subsamples, ...
- Weka supports several schemes for **attribute selection**. Both filter methods and wrapper methods are supported. Among the provided filter methods are the chi-squared method, the information gain and gain ratio measures, the performance of the OneR classifier based on each single attribute, **PCA**, the relief-F method (based on distances between sampled instances), ... In wrapper methods the feature subset is evaluated using the actual classifier that is going to be used for classification. This is obviously quite time-consuming. When searching for an optimal feature subset, a search strategy has to be selected. Possibilities are: best-first, exhaustive search, forward selection, ranking, genetic search, random search, ...
- **Visualisation:** Weka provides limited visualisation possibilities. there are maximum three dimensions: 2 axis and one overlay colour.

Advantages: The obvious advantage of a package like Weka is that **a whole range of data preparation, feature selection and data mining algorithms are integrated**. This means that only one data format is needed, and trying out and comparing different approaches becomes really easy. The package also comes with **a GUI**, which should make it easier to use.

Disadvantages: Probably the most important disadvantage of data mining suites like this is that **they do not implement the newest techniques**. For example the MLP implemented has a very basic training algorithm (backprop with momentum), and the SVM only uses polynomial kernels, and does not support numeric estimation. Therefore, it will be necessary to combine WEKA with some of the other tools like Netlab or SVM_torch. Another important disadvantage arises from the fact that the software is for free: **the documentation for the GUI is quite limited**. As the software is constantly growing, the documentation is not up to date with everything either (the most up to date and complete information about algorithm options can be obtained by using the -h option in the command line interface).

A third possible problem is scaling. For difficult tasks on large datasets, the running time can become quite long, and java sometimes gives an OutOfMemory error. This problem can be reduced by using the '-mx*x*' option when calling java, where *x* is memory size (*e.g.*, '50 m'). For large datasets it will always be necessary to reduce the size to be able to work within reasonable time limits. A fourth problem is that **the GUI does not implement all the possible options**. Things that could be very useful, like scoring of a test set, are not provided in the GUI, but can be called from the command line interface. So sometimes it will be necessary to switch between GUI and command line. Finally, **the data preparation and visualisation techniques offered might not be enough**. Most of them are very useful, but I think in most data mining tasks you will need more to get to know the data well and to get it in the right format.

Bow

General Description: Bow is a library of C code for statistical text analysis, language modeling and information retrieval. Together with this library, four executable programs based on it are distributed. They are Rainbow (for document classification), Arrow and Archer (for document retrieval) and Crossbow (for document clustering).

Functionalities:

- **Rainbow**: Rainbow is the front-end to the library that supports text classification. It is the best documented of the four, and is the most useful for the projects. It has the following functionalities:
  - (*a*) **Text preprocessing**: This supports the conversion of texts into bag-of-words format. Different options allow to choose a sparse or a full matrix model, to display a count or a binary present/absent flag, and to include the word names or not. Also, feature selection methods can be used, based on word frequency, word-document counts or information gain.
  - (*b*) **Naive Bayes classification**
  - (*c*) **kNN classification**
  - (*d*) **Support Vector Machine classification**
  - (*e*) **Tfidf classification**
  - (*f*) **Expectency Maximisation**
- **Crossbow:** Crossbow supports text clustering. No documentation is available for it yet. Some indications of how to use it can be found in the help file.
- **Arrow:** Arrow does document retrieval based on tfidf. There is a little bit of information available in the README file.
- **Archer:** Archer also does document retrieval. No documentation is available yet though.
- **Evaluation:** Especially the **text preprocessing functionalities** are very useful for text related projects. Once the conversion to bag-of-word format is done, you have the option to use the classification and clustering features of the bow package, or to export the matrix to other data mining programs. The classification functionalities of Rainbow give **very fast and accurate results**.

Netlab

General description: Netlab is a toolbox of matlab functions and scripts, originally based on Bishop's *Neural networks for pattern recognition* book. They provide matlab implementations of some of the newest machine learning algorithms.

Functionalities: This is a list of the functionalities supported by Netlab.

- PCA
- Mixtures of probabilistic PCA
- Gaussian mixture model with EM training algorithm
- Linear and logistic regression with IRLS training algorithm
- Multi-layer perceptron with linear, logistic and softmax outputs and appropriate error functions
- Radial basis function (RBF) networks with both Gaussian and non-local basis functions
- Optimisers, including quasi-Newton methods, conjugate gradients and scaled conjugate gradients
- Multi-layer perceptron with Gaussian mixture outputs (mixture density networks)
- Gaussian prior distributions over parameters for the MLP, RBF and GLM including multiple hyper-parameters
- Laplace approximation framework for Bayesian inference (evidence procedure)
- Automatic Relevance Determination (ARD) for input selection
- Markov chain Monte-Carlo including simple Metropolis and hybrid Monte-Carlo
- K-nearest neighbour classifier
- K-means clustering
- Generative Topographic Map
- Neuroscale topographic projection
- Gaussian Processes
- Hinton diagrams for network weights
- Self-organising map

Advantages: Working with netlab surely offers some important advantages. Compared to a package like Weka, the greatest advantage is probably that **the algorithms implemented are up to date** with the newest developments in the field. Also, using the functions assumes quite some **knowledge of the theory behind the algorithms**, so at least the data mining analyst knows what is going on. Finally, the **integration in matlab** gives some important advantages: all the matrix calculation and visualisation functions are available, which can be very useful for data preparation and exploration, and analysis of results afterwards. Also, Matlab makes scripting is possible, which should increase the efficiency of the analysis.

Disadvantages: Disadvantages are that **some important data preprocessing functionalities are missing**: dealing with missing data, feature selection, ... Another problem is that matlab **can only work with numeric data**, so categorical data will have to be converted in advance, and techniques like apriori and decision trees cannot be implemented.

SVMTorch

General Description: SVMTorch is an implementation of Vapnik's Support Vector Machine that works both for classification and regression problems, and that has been specifically tailored for large-scale problems (such as more than 20000 examples, even for input dimensions higher than 100).

Functionalities: SVMTorch implements support vector classification and regression. The data have to be floating point values. It is possible to do classification for multiple classes: then a separate classifier is built for every class. A small user guide for SVMTorch is available on-line.

Evaluation: SVMTorch offers many options for the use of SVM's. Linear, polynomial, gaussian, sigmoidal and even user-defined kernels are possible. Regression as well as classification can be done, and there is support for multiple classes. This extended functionality is in sharp contrast with the limited SVM support in Weka. Also, SVMTorch is very fast and scales well to large amounts of data. A disadvantage of using SVMTorch is that a separate data format is needed.

XGobi

General Description: This is an interactive visualisation tool for high-dimensional data. It supports 2-D plots, 3-D rotations, scaling of axes, linked brushing (allowing to color certain points, which will then stand out in different views of the data), and much more. A good overview of the functionalities of xgobi is given in *XGobi: Interactive Dynamic Data Visualisation in the X Window System (1998)* by D. F. Swayne, D. Cook and A. Buja.

Evaluation: This is definitely a useful tool when you do exploratory data analysis. Weka has a visualisation functionality, and also Matlab allows to plot data, but they don't offer the dynamic and interactive features of XGobi.

C4.5

General Description: This is software distributed by Quinlan to build decision trees using his c4.5 algorithm. Information about how to use it can be found by consulting the man pages (type 'man c4.5'). More information about the working of c4.5 can be found in Mitchell's Machine Learning book.

Evaluation: The c4.5 algorithm is also embedded in Weka, so it makes sense to use it there if you are using the other Weka functionalities. The separate c4.5 program seems to go a little faster than the java implementation in Weka, however, so for large files, it might be useful to use this.

Cluster

General Description: This program supports hierarchical clustering and PCA. Information about how to use it can be found in the man pages (type 'man 1 cluster').

Evaluation: Hierarchical clustering is not provided in any of the other software packages presented above, so if you want to do that, you'll have to use this program. It is quite slow though, so don't try to use it on large amounts of data. The pca functionality is also available in netlab and in weka, so you can use it there.

Question Bank

APPENDIX C

1. *What are the uses of statistics in data mining?*

Ans. Statistics is used to

- to estimate the complexity of a data mining problem;
- suggest which data mining techniques are most likely to be successful; and
- identify data fields that contain the most "surface information".

2. *What is the main goal of statistics?*

Ans. The basic goal of statistics is to extend knowledge about a subset of a collection to the entire collection.

3. *What are the factors to be considered while selecting the sample in statistics?*

Ans. The sample should be

- Large enough to be representative of the population.
- small enough to be manageable.
- accessible to the sampler.
- free of bias.

4. *Name some advanced database systems.*

Ans. Object-oriented databases, Object-relational databases.

5. *Name some specific application oriented databases.*

Ans. Spatial databases, Time-series databases, Text databases and multimedia databases.

6. *Define Relational databases.*

Ans. A relational database is a collection of tables, each of which is assigned a unique name. Each table consists of a set of attributes (columns or fields) and usually stores a large set of tuples (records or rows).Each tuple in a relational table represents an object identified by a unique key and described by a set of attribute values.

7. *Define Transactional Databases.*

Ans. A transactional database consists of a file where each record represents a transaction. A transaction typically includes a unique transaction identity number (trans_ID), and a list of the items making up the transaction.

8. *Define Spatial Databases.*

Ans. Spatial databases contain spatial-related information. Such databases include geographic (map) databases, VLSI chip design databases, and medical and satellite image databases. Spatial data may be represented in raster format, consisting of n-dimensional bit maps or pixel maps.

9. *What is Temporal Database?*

Ans. Temporal database store time related data .It usually stores relational data that include time related attributes. These attributes may involve several time stamps, each having different semantics.

10. *What is Time-Series databases?*

Ans. A Time-Series database stores sequences of values that change with time, such as data collected regarding the stock exchange.

11. *What is Legacy database?*

Ans. A Legacy database is a group of heterogeneous databases that combines different kinds of data systems, such as relational or object-oriented databases, hierarchical databases, network databases, spread sheets, multimedia databases or file systems.

12. *What is learning?*

Ans. Learning denotes changes in the system that enables the system to do the same task more efficiently the next time. Learning is making useful changes or modifying what is being experienced.

13. *Why machine learning is done?*

Ans. To understand and improve the efficiency of human learning, To discover new things or structure that is unknown to human beings, To fill in skeletal or computer specifications about a domain.

14. *Give the components of a learning system.*

Ans. The components of learning system are:

- Critic
- Sensors
- Learning Element
- Performance Element
- Effectors
- Problem generators.

15. *Give some of the factors for evaluating performance of a learning algorithm.*

Ans. Predictive accuracy of a classifier, Speed of a learner, speed of a classifier, Space requirements.

16. *What are the steps in the data mining process?*

Ans. Data cleaning, Data integration, Data selection, Data transformation, Data mining, Pattern evaluation, Knowledge representation.

17. *Define data cleaning.*

Ans. Data cleaning means removing the inconsistent data or noise and collecting necessary information.

18. *Define data mining.*

Ans. Data mining is a process of extracting or mining knowledge from huge amount of data.

19. *Define pattern evaluation.*

Ans. Pattern evaluation is used to identify the truly interesting patterns representing knowledge based on some interesting measures.

20. *Define knowledge representation.*

Ans. Knowledge representation techniques are used to present the mined knowledge to the user.

21. *What is Visualization?*

Ans. Visualization is for depiction of data and to gain intuition about data being observed. It assists the analysts in selecting display formats, viewer perspectives and data representation schema.

22. *Name some conventional visualization techniques.*

Ans. Histogram, Relationship tree, Bar charts, Pie charts, Tables etc.

23. *Give the features included in modern visualization techniques.*

Ans. Morphing, Animation, Multiple simultaneous data views, Drill-Down and Hyperlinks to related data source.

24. *Define conventional visualization.*

Ans. Conventional visualization depicts information about a population and not the population data itself.

25. *Define Spatial Visualzation.*

Ans. Spatial visualization depicts actual members of the population in their feature space.

26. *What is Descriptive and predictive data mining?*

Ans. Descriptive data mining describes the data set in a concise and summarative manner and presents interesting general properties of the data.

Predictive data mining analyzes the data in order to construct one or set of models and attempts to predict the behaviour of new data sets.

27. *What is Data Generalization?*

Ans. It is a process that abstracts a large set of task-relevant data in a database from a relatively low conceptual to higher conceptual levels. The approaches for Generalization are

1. Data cube approach
2. Attribute-oriented induction approach

28. *Define Attribute Oriented Induction.*

Ans. These method collects the task-relevant data using a relational database query and then perform generalization based on the examination in the relevant set of data.

29. *What is Jack Knife?*

Ans. It's a bias reduction tool for eliminating low order bias from an estimator. The essence of the procedure is to replace the original 'n' observations by 'n' more correlated estimates of the quantity of interest. These are obtained by systematically leaving out one or more observations and re-computing the estimator.

30. *What is boot strap?*

Ans. An interpretation of the jack knife is that the construction of pseudo value is based on repeatedly and systematically sampling with out replacement from the data at hand. This lead to generalized concept to repeated sampling with replacement called boot strap.

31. *View of statistical approach.*

Ans. Statistical method is interested in interpreting the model. It may sacrifice some performance to be able to extract meaning from the model structure. If accuracy is acceptable then the reason that a model can be decomposed in to revealing parts is often more useful than a 'black box' system, especially during early stages of investigation and design cycle.

32. *What are the assumptions of statistical analysis?*

Ans. The assumptions are Residuals, Diagnostics and Parameter Covariance.

33. *What is the use of Probabilistic graphical model?*

Ans. Probabilistic graphical model are a frame work for structuring, representation and decomposing a problem using the notation of conditional independence.

34. *What is the importance of Probabilistic graphical model?*

Ans. They are a lucid representation for a variety of problems, allowing key dependencies within a problem to be expressed and irrelevancies to be ignored. It performs problem formulation and decomposition. It helps in designing a learning algorithm. It identifies valuable knowledge. It generates explanation.

35. *Define Deterministic models.*

Ans. Deterministic models, which takes no account of random variables, but gives precise, fixed reproducible output.

36. *Define Systems and Models.*

Ans. System is a collection of interrelated objects and Model is a description of a system. Models are abstract, and conceptually simple.

37. *How do you choose the best model?*

Ans. All things being equal, the smallest model that explains the observations and fits the objectives should be accepted. In reality, the smallest means the model should optimizes a certain scoring function (*e.g.*, Least nodes, most robust, least assumptions)

38. *What are the Principles of Qualitative Formulation?*

Ans. Model Simplification, Minimize state variables, Convert a variable into a constant aggregate state variable, Make stronger assumptions, Remove temporal complexity, Remove spatial complexity.

39. *What are the general properties of Boolean Networks?*

Ans. Fixed topology, Dynamic synchronous Node States, Gate function, Synergetic behavior

40. *What is clustering?*

Ans. Clustering is the process of grouping the data into classes or clusters so that objects within a cluster have high similarity in comparison to one another, but are very dissimilar to objects in other clusters.

41. *What are the requirements of clustering?*

Ans. Scalability, Ability to deal with different types of attributes, Ability to deal with noisy data, Minimal requirements for domain knowledge to determine input parameters, Constraint based clustering, Interpretability and usability.

42. *State the categories of clustering methods.*

Ans. Partitioning methods, Hierarchical methods, Density based methods, Grid based methods, Model based methods

43. *What is linear regression?*

Ans. In linear regression data are modeled using a straight line. Linear regression is the simplest form of regression. Bivariate linear regression models a random variable Y called response variable as a linear function of another random variable X, called a predictor variable.

$$Y = a + b\,X.$$

44. *State the types of linear model and state its use.*

Ans. Generalized linear model represent the theoretical foundation on which linear regression can be applied to the modeling of categorical response variables. The types of generalized linear model are

(*i*) Logistic regression

(*ii*) Poisson regression

45. *What are the goals of Time series analysis?*

Ans. Finding Patterns in the data and Predicting future values

46. *What is smoothing?*

Ans. Smoothing is an approach that is used to remove nonsystematic behaviors found in a time series. It can be used to detect trends in time series.

47. *What is lag?*

Ans. The time difference between related items is referred to as lag.

48. *Write the preprocessing steps that may be applied to the data for classification and prediction.*

Ans. Data Cleaning, Relevance Analysis and Data Transformation.

49. *Define Data Classification.*

Ans. It is a two-step process. In the first step, a model is built describing a pre-determined set of data classes or concepts. The model is constructed by analyzing database tuples described by attributes. In the second step the model is used for classification.

50. *What are Bayesian Classifiers?*

Ans. Bayesian Classifiers are statistical classifiers. They can predict class member-ship probabilities, such as the probability that a given sample belongs to a particular class.

51. *Describe the two common approaches to tree pruning.*

Ans. In the prepruning approach, a tree is "pruned" by halting its construction early. The second approach, postpruning, removes branches from a "fully grown" tree. A tree node is pruned by removing its branches.

52. *What is a "decision tree"?*

Ans. It is a flow-chart like tree structure, where each internal node denotes a test on an attribute, each branch represents an outcome of the test, and leaf nodes represent classes or class distributions. Decision tree is a predictive model. Each branch of the tree is a classification question and leaves of the tree are partition of the dataset with their classification.

53. *What do you meant by concept hierarchies?*

Ans. A concept hierarchy defines a sequence of mappings from a set of low-level concepts to higher-level, more general concepts. Concept hierarchies allow specialization, or drilling down, where by concept values are replaced by lower-level concepts.

54. *Where are decision trees mainly used?*

Ans. Used for exploration of dataset and business problems, Data preprocessing for other predictive analysis and Statisticians use decision trees for exploratory analysis

55. *How will you solve a classification problem using decision trees?*

Ans. Various methods are:

(*a*) Decision tree induction: Construct a decision tree using training data.

(*b*) For each ti ∈∈∈ D apply the decision tree to determine its class ti-tuple D-Database.

56. *What is decision tree pruning?*

Ans. Once a tree is constructed, some modification to the tree might be needed to improve the performance of the tree during classification phase. The pruning phase might remove redundant comparisons or remove subtrees to achieve better performance.

57. *Explain ID3.*

Ans. ID3 is algorithm used to build decision tree. The following steps are followed to built a decision tree.

(*a*) Chooses splitting attribute with highest information gain.

(*b*) Split should reduce the amount of information needed by large amount.

58. *What is Association rule?*

Ans. Association rule finds interesting association or correlation relationships among a large set of data items which is used for decision-making processes. Association rules analyzes buying patterns that are frequently associated or purchased together.

59. *Define support.*

Ans. Support is the ratio of the number of transactions that include all items in the antecedent and consequent parts of the rule to the total number of transactions. Support is an association rule interestingness measure.

60. *Define Confidence.*

Ans. Confidence is the ratio of the number of transactions that include all items in the consequent as well as antecedent to the number of transactions that include all items in antecedent. Confidence is an association rule interestingness measure.

61. *How are association rules mined from large databases?*

Ans. Association rule mining is a two-step process.

- Find all frequent itemsets.
- Generate strong association rules from the frequent itemsets.

62. *What is the classification of association rules based on various criteria?*

Ans. 1. Based on the types of values handled in the rule.

(*a*) Boolean Association rule.

(*b*) Quantitative Association rule.

2. Based on the dimensions of data involved in the rule.

(*a*) Single Dimensional Association rule.

(*b*) Multi Dimensional Association rule.

3. Based on the levels of abstractions involved in the rule.

(*a*) Single level Association rule.

(*b*) Multi level Association rule.

4. Based on various extensions to association mining.

(*a*) Maxpatterns.

(*b*) Frequent closed itemsets.

63. *What is Apriori algorithm?*

Ans. Apriori algorithm is an influential algorithm for mining frequent itemsets for Boolean association rules using prior knowledge. Apriori algorithm uses prior knowledge of frequent itemset properties and it employs an iterative approach known as level-wise search where k-itemsets are used to explore $(k + 1)$-itemsets.

64. *What are the advantages of Dimensional modelling?*

Ans. Ease of use, High performance, Predictable, standard framework, Understandable, Extensible to accommodate unexpected new data elements and new design decisions

65. *Define Dimensional Modelling?*

Ans. Dimensional modelling is a logical design technique that seeks to present the data in a standard framework that intuitive and allows for high-performance access.It is inherently dimensional and adheres to a discipline that uses the relational model with some important restrictions.

66. *What comprises of a dimensional model?*

Ans. Dimensional model is composed of one table with a multipart key called fact table and a set of smaller tables called dimension table. Each dimension table has a single part primary key that corresponds exactly to one of the components of multipart key in the fact table.

67. *Define a data mart?*

Ans. Data mart is a pragmatic collection of related facts, but does not have to be exhaustive or exclusive. A data mart is both a kind of subject area and an application. Data mart is a collection of numeric facts.

68. *What are the advantages of a data modelling tool?*

Ans. Integrates the data warehouse model with other corporate data models, Helps assure consistency in naming, Creates good documentation in a variety of useful formats, Provides a reasonably intuitive user interface for entering comments about objects.

69. *What is data warehouse performance issue?*

Ans. The performance of a data warehouse is largely a function of the quantity and type of data stored within a database and the query/data loading work load placed upon the system.

70. *What are the types of performance issue?*

Ans. Capacity planning for the data warehouse, data placement techniques within a data warehouse, Application Performance Techniques, Monitoring the Data Warehouse.

71. *What is Data Inconsistency Cleaning?*

Ans. This can be summarized as the process of cleaning up the small inconsistencies that introduce themselves into the data. Examples include duplicate keys and unreferenced foreign keys.

72. *What is Column Level Cleaning?*

Ans. This involved checking the contents of each column field and ensuring it conforms to a set of valid values. For example convert EBCDIC to ASCII.

73. *What is Bottleneck Detection?*

Ans. Bottleneck Detection is the process whereby the database administrator will detect for what reason the database performance level has reached a plateau. To increase the performance from this plateau may require the addition of more hardware resources or reconfiguration of the system software (O/S, RDBMS or Application).

74. *What is back room Meta data?*

Ans. Back room Meta data is process related, and it guides the extraction, cleaning and loading process.

75. *What is Front Room Meta data?*

Ans. It is more descriptive and it helps query tools and report writers function smoothly.

76. *What are the specifications of source system Meta data?*

Ans. Repositories, Source Scheme, Copy Books, Spread Sheet Sources and Lotus notes database

77. *What is active Meta data?*

Ans. Active Meta data is Meta data that drives a process rather than documents.

78. *What is Meat data catalogue?*

Ans. Meat data catalogue is a single common storage point for information that drives the entire warehouse process.

79. *Why do you need data warehouse life cycle process?*

Ans. Data warehouse life cycle approach is essential because it ensures that the project pieces are brought together in the right order and at the right time.

80. *What are the steps in the life cycle approach?*

Ans. Project Planning, Business Requirements definition, Data track: Dimensional modeling, Physical Design, Data Staging Design and Development, Technology track: Technical Architecture design, Product Selection and Installation, Application track: End user Application Specification, End user Application Development, Deployment, Maintenance and Growth and Project Management

81. *What are the merits of Data Warehouse?*

Ans. The merits are:

- Ability to make effective decisions from database
- Better analysis of data and decision support
- Discover trends and correlations that benefits business
- Handle huge amount of data.

82. *What are the characteristics of data warehouse?*

Ans. Separate, Available, Integrated, Subject Oriented, Not Dynamic, Consistency, Iterative Development, Aggregation Performance

83. *List some of the Data Warehouse tools.*

Ans. Various tools are:

- OLAP (OnLine Analytic Processing)
- ROLAP (Relational OLAP)
- End User Data Access tool
- Ad Hoc Query tool
- Data Transformation services
- Replication

84. *Explain OLAP.*

Ans. The general activity of querying and presenting text and number data from Data Warehouses, as well as a specifically dimensional style of querying and presenting that is exemplified by a number of "OLAP Vendors" .The OLAP vendors technology is non-relational and is almost always biased on an explicit multidimensional cube of data. OLAP databases are also known as multidimensional cube of databases.

85. *Explain ROLAP.*

Ans. ROLAP is a set of user interfaces and applications that give a relational database a dimensional flavour. ROLAP stands for Relational Online Analytic Processing.

86. *Explain End User Data Access tool.*

Ans. End User Data Access tool is a client of the data warehouse. In a relational data warehouse, such a client maintains a session with the presentation server, sending a stream of separate SQL requests to the server. Eventually the end user data access tool is done with the SQL session and turns around to present a screen of data or a report, a graph, or some other higher form of analysis to the user. An end user data access tool can be as simple as an Ad Hoc query tool or can be complex as a sophisticated data mining or modeling application.

87. *Explain Ad Hoc query tool.*

Ans. A specific kind of end user data access tool that invites the user to form their own queries by directly manipulating relational tables and their joins. Ad Hoc query tools, as powerful as they are, can only be effectively used and understood by about 10% of all the potential end users of a data warehouse.

88. *Name some of the data mining applications.*

Ans. The applications are:

- Data mining for Biomedical and DNA data analysis.
- Data mining for Financial data analysis.
- Data mining for the Retail industry.
- Data mining for the Telecommunication industry.

89. *What is the contribution of data mining to DNA analysis.*

Ans.

- Semantic integration of heterogeneous, distributed genome databases.
- Similarity search and comparison among DNA sequences.
- Association analysis: identification of co-occuring gene sequences.
- Path analysis: linking genes to different stages of disease development.
- Visualization tools and genetic data analysis.

90. *Name some examples of data mining in retail industry.*

Ans.

- Design and construction of data warehouses based on the benefits of data mining.
- Multidimensional analysis of sales, customers, products, time and region.
- Analysis of the effectiveness of sales campaigns.
- Customer retention-analysis of customer loyalty.
- Purchase recommendation and cross-reference of item.

91. *Name some of the data mining applications.*

Ans.

- Data mining for Biomedical and DNA data analysis.
- Data mining for Financial data analysis.
- Data mining for the Retail industry.
- Data mining for the Telecommunication industry.

92. *What is the difference between "supervised" and unsupervised" learning scheme?*

Ans. In data mining during classification the class label of each training sample is provided, this type of training is called supervised learning (*i.e.*) the learning of the model is supervised in that it is told to which class each training sample belongs. For example, Classification.

In unsupervised learning the class label of each training sample is not known and the member or set of classes to be learned may not be known in advance. For example, Clustering.

93. *Discuss the importance of similarity metric clustering. Why is it difficult to handle categorical data for clustering?*

Ans. The process of grouping a set of physical or abstract objects into classes of similar objects is called clustering. Similarity metric is important because it is used for outlier detection. The clustering algorithm which is main memory based can operate only on the following two data structures namely,

(A) Data matrix

(B) Dissimilarity matrix

So it is difficult to handle categorical data.

94. *Mention atleast 3 advantages of Bayesian Networks for data analysis. Explain each one.*

Ans. (*a*) Bayesian Network is a graphical representation of unknown knowledge that is easy to construct and interpret.

(*b*) The representation has formal probabilistic semantics, making it suitable for statistical manipulation.

(*c*) The representation is used for encoding uncertain expert knowledge in expert systems.

95. *Why do we need to prune a decision tree? Why should we use a separate pruning data set instead of pruning the tree with the training database?*

Ans. When a decision tree is built, many of the branches will reflect animation in the training data due to noise or outliers. Tree pruning methods are needed to address this problem of overfitting the data.

96. *Explain the various OLAP operations.*

Ans. (*a*) ***Roll-up:*** The roll-up operation performs aggregation on a data cube, either by climbing up a concept hierarchy for a dimension.

(*b*) ***Drill-down:*** It is the reverse of roll-up. It navigates from less detailed data to more detailed data.

(*c*) ***Slice:*** Performs a selection on one dimension of the given cube, resulting in a subcube.

97. *Discuss the concepts of frequent itemset, support and confidence.*

Ans. A set of items is referred to as itemset. An itemset that contains k items is called k-itemset. An itemset that satisfies minimum support is referred to as frequent itemset.

Support is the ratio of the number of transactions that include all items in the antecedent and consequent parts of the rule to the total number of transactions.

Confidence is the ratio of the number of transactions that include all items in the consequent as well as antecedent to the number of transactions that include all items in antecedent.

98. *Why is data quality so important in a data warehouse environment?*

Ans. Data quality is important in a data warehouse environment to facilitate decision-making. In order to support decision-making, the stored data should provide information from a historical perspective and in a summarized manner.

99. *How can data visualization help in decision-making?*

Ans. Data visualization helps the analyst gain intuition about the data being observed. Visualization applications frequently assist thc analyst in selecting display formats, viewer perspective and data representation schemas that faster deep intuitive understanding thus facilitating decision-making.

100. *What do you mean by high performance data mining?*

Ans. Data mining refers to extracting or mining knowledge. It involves an integration of techniques from multiple disciplines like database technology, statistics, machine learning, neural networks, etc. When it involves techniques from high performance computing it is referred as high performance data mining.

101. *What are the various data mining issues?*

Ans. Knowledge Mining, User interaction, Performance, Diversity in datatypes

102. *What are the data mining functionalities?*

Ans. The data mining functionalities are:

- Concept class description
- Association analysis
- Classification and prediction
- Cluster Analysis
- Outlier Analysis

103. *What are the different types of data repositories on which mining can be performed?*

Ans. The different types of data repositories on which mining can be performed are:

- Relational Databases
- DataWarehouses
- Transactional Databases
- Advanced Databases
- Flat files
- World Wide Web

104. *What is Data Mining? What are the steps in Knowledge Discovery?*

Ans. Data mining refers to extracting or mining knowledge from large amount of data. The steps in knowledge discovery are:

- Data cleaning
- Data integration
- Data selection
- Data transformation
- Data mining
- Pattern Evolution
- Knowledge Discovery.

105. *What are the data pre-processing techniques?*

Ans. The data preprocessing techniques are:

- Data Cleaning
- Data integration
- Data transformation
- Data reduction

106. *What are the smoothing Techniques?*

Ans.

- Binning
- Clustering
- Regression

107. *What are parametric methods and non-parametric methods of reduction?*

Ans. Parametric Methods:

- Regression Model
- Log linear Model

Non-Parametric Methods

- Sampling
- Histogram
- Clustering

108. *Explain Data Discretization and Concept Hierarchy Generation.*

Ans. Discretization and concept hierarchy generation for numerical data:

- Segmentation by natural partitioning
- Binning
- Histogram Analysis
- Cluster Analysis

109. *What are Data mining Primitives?*

Ans. There are 5 Data mining Primitives. They are:

- Task relevant data
- Kinds of knowledge to be mined
- Concept Hierarchies
- Interesting Measures
- Knowledge Presentation and Visualization Technique to be used for Discovery patterns.

110. *Explain various classification methods?*

Ans.

- K-nearest neighbour
- Case-Based reasoning
- Genetics
- Fuzzy logic

111. *Briefly explain about Data Mining Application.*

Ans. Various applications are:

- Financial Data Analysis
- Retail Industry
- Telecommunication Industry
- Biological Data Analysis
- Scientific Application

112. *What is the need for Data Processing?*

Ans. To get the required information from huge, incomplete, noisy and inconsistent set of data it is necessary to use data processing.

113. *What is Data Cleaning?*

Ans. **Data cleaning** is a procedure to "clean" the data by filling in missing values, smoothing noisy data, identifying or removing outliers, and resolving inconsistencies

114. *What is Data Integration?*

Ans. Integrating multiple databases, data cubes, or files, that is, **data integration.**

115. *What is Data Transformation?*

Ans. **Data transformation** operations, such as normalization and aggregation, are additional data preprocessing procedures that would contribute toward the success of the mining process.

116. *What is Data Reduction?*

Ans. **Data reduction** obtains a reduced representation of the data set that is much smaller in volume, yet produces the same (or almost the same) analytical results.

■■■

CHAPTER 1

Index

Symbols

A

B

C

D

E

F

G

H

I

K

L

M

N

U

V

W

Z

Win Prizes

ATTENTION READERS

Win Prizes

We request for your frank assessment regarding various aspects of the book as given below:

DATA MINING AND DATA WAREHOUSING

BY : Kanika Lakhani Chaudhary and Gaurav Girdhar (UTU)

Please Fill up the Sheet in Neat and Clean Writing

(i) What topics of your syllabus, important for your examination point of view, are not covered in this book ?

....

....

....

....

(ii) What are the chapters wherein the treatment of the subject matter is not systematic or organised or updated ?

....

....

....

....

(iii) Have you come across misprints/mistakes/factual inaccuracies in the book ? Please specify the chapters and the page numbers.

....

....

....

....

(iv) What other books on the same subject have you found/heard better than the present book ? Please specify in terms of price and quality.

....

....

....

....

(v) Further suggestions and comments for the improvement of this book.

....

....

....

....

(PS : If need be please attach a separate sheet to write your views)

OTHER DETAILS

(i) Who recommended you this book ? Please tick (✓) in the box near the option relevant to you.

☐ Teacher ☐ Friends ☐ Bookseller

....

....

....

....

(ii) Name of the recommending teacher, his designation and address:

....

....

....

....

(iii) Name and address of the bookseller you purchased the book from :

....

....

....

....

(iv) Name and address of your institution (Please mention the University or Board as the case may be)

....

....

....

....

(v) Your name and complete address :

....

....

....

....

(vi) Course you are enrolled in :

....

....

....

....

The best assessment will be awarded each month : The award will be in the form of our publications as decided by the Editorial Board.

Please mail the filled up coupon at your earliest to :

S.K. KATARIA & SONS®
4760-61/23, Ansari Road, Daryaganj, New Delhi-110002
☎ +91-11-23243489, 23269324; Telefax: +91-11-23243489
e-mail: katariabooks@yahoo.com; katariabook@gmail.com
Website : www.skkatariaandsons.com